Nanotechnology and Functional Materials for Engineers

Nanotechnology and Functional Materials for Engineers

Yaser Dahman
Ryerson University, Toronto, Canada

ELSEVIER   elsevier.com

Elsevier
Radarweg 29, PO Box 211, 1000 AE Amsterdam, Netherlands
The Boulevard, Langford Lane, Kidlington, Oxford OX5 1GB, United Kingdom
50 Hampshire Street, 5th Floor, Cambridge, MA 02139, United States

Notices
Knowledge and best practice in this field are constantly changing. As new research and experience broaden our understanding, changes in research methods, professional practices, or medical treatment may become necessary. Practitioners and researchers must always rely on their own experience and knowledge in evaluating and using any information, methods, compounds, or experiments described herein. In using such information or methods they should be mindful of their own safety and the safety of others, including parties for whom they have a professional responsibility.

To the fullest extent of the law, neither the Publisher nor the authors, contributors, or editors, assume any liability for any injury and/or damage to persons or property as a matter of products liability, negligence or otherwise, or from any use or operation of any methods, products, instructions, or ideas contained in the material herein.

British Library Cataloguing-in-Publication Data
A catalogue record for this book is available from the British Library

Library of Congress Cataloging-in-Publication Data
A catalog record for this book is available from the Library of Congress

ISBN: 978-0-323-51256-5

For Information on all Elsevier publications
visit our website at https://www.elsevier.com

Working together
to grow libraries in
developing countries

www.elsevier.com • www.bookaid.org

Publisher: Matthew Deans
Acquisition Editor: Simon Holt
Editorial Project Manager: Sabrina Webber
Production Project Manager: Susan Li
Designer: Greg Harris

Typeset by MPS Limited, Chennai, India

Contents

Biography

Dr. Yaser Dahman is a professor in the Department of Chemical Engineering at Ryerson University in Toronto, Canada. His credential includes doctoral degree in Chemical and Biochemical Engineering and Master of Business Administration. He commissioned and served as the director of the Nanocomposites and Biomaterials Engineering Laboratory and Centre of Green Research Technology, both at Ryerson University. He also served as the director of the Chemical Engineering undergraduate program and Engineering Innovation and Entrepreneurship master program at Ryerson. Dr. Dahman chaired the "4th International Conference on Nanotechnology: Fundamentals and Applications" that was held at Ryerson University in 2013.

Preface

The textbook presents a broad scoop of the entire field of nanotechnology, focusing on key essentials, and delivers examples of applications in various fields. The textbook is relevant academically for advanced undergraduate and graduate students in majors such as Chemical Engineering, Material Science, Electrical Engineering, Electronics, Biomedical Engineering, Bioengineering, Nanoscience, since it would assist in understanding the main concepts of nanotechnology. Moreover, researchers in nanoscience and nanotechnology and, in particular, college and university professors of faculties other than engineering (e.g., specialists working in the fields of surface science, chemistry, solid-state physics, materials science, polymer science, colloid science, environmental science, pharmacy, biotechnology, drug delivery, etc.) will benefit from this book. The book would be an addition to the science and engineering libraries, which can be preferred as a textbook or reference book.

In general this book comprehends and illustrates nanoelectronics and electrooptic nanotechnology, covering the physics, nanostructures, and nanodevices. Additionally the book shows specific topics related to nanotechnology, such as nanosensors, smart nanomaterials, nanopolymers, nanotubes, synthesis and characteristics of nanostructures, production methods, and applications, all discussed in a simple and comprehensive manner. Topics were arranged systematically. Furthermore, the information presented will surely build a cognitive bridge between the specific experiments and up-to-date research and applications of nanostructures. The mix of knowledge presented by introducing nanomaterial concepts, characterization techniques, and fields of applications chapter-by-chapter in relation to the specific properties gained in products when these nanomaterials are applied. Professionals outside of Materials Science can find a quick reference to the applications and metrology of nanostructures. This book has also the potential to provide more information qualitatively and quantitatively. For example, although the CRC book gives a thorough description of imaging and spectroscopy techniques, AFM and STM examples give a good description of how to analyze the images obtained.

In this book, Chapter 1, An Introduction to Nanotechnology, gives a comprehensive introduction to the topic of nanotechnology, including the current definition of nanotechnology, its history, applications, recent researches, and future perspectives. Chapter 1, An Introduction to Nanotechnology, also gives a good summary of the topics discussed in other chapters, such as methodologies for characterization, "smart" nanomaterials, nanosensors, etc. Chapter 2, Generic Methodologies for Characterization, presents the instrumentations and methodologies used to study the characteristics of nanoparticles through the use of different apparatus and techniques. These techniques include: scanning probes (STM, AFM, MFM), electron microscopy, transmission electron microscopy, diffraction (XRD), spectroscopy, surface analysis and depth profiling (AES). Chapter 3, Smart Nanomaterials, summarizes the different types of nanomaterials based on the size, mechanism of response, and synthesis technique. Moreover, various applications of these nanomaterials are explored at the end of the chapter. Chapter 4, Nanosensors, discusses the types of nanosensors and their uses in nanotechnology as well as major methods of manufacturing nanosensors, and their possible application. Chapter 5, Nanoparticles, looks at the synthesis of nanoparticles, such as magnetic nanoparticles, and their applications in various fields such as biomedical and biotechnological. Chapter 6, Nanopolymers, gives a comprehensive summary about polymeric nanofibers, including their fabrication (electrospinning), their environmental (fuel cells, heavy metal ion removal and solar energy) and biomedical (wound healing and biomaterials) application.

This chapter also looks at the application of nanofibers as biosensors. Chapter 7, Nanotubes, discusses about the types of carbon nanotubes and their history, properties (mechanical, electrical, and thermal), production methods (arc discharge, laser ablation, and chemical vapors deposition), purification, and application. Chapter 8, Nanoshells, gives a comprehensive introduction to nanoshells and their application in industry and medicine. This chapter also reviews the types of nanoshells and their properties (optical, luminescence, thermal, etc.). Chapter 9, Electronic and Electro-optic Nanotechnology, gives a good breakdown of electronic and electrooptic nanotechnology. It looks at the theory of electronic and electrooptic nanotechnology and the future perspective of this emerging technology. Chapter 10, Self-Assembling Nanostructures, illustrates the principles of self-assembling nanostructures as well as their methods of preparation and future perspectives. Chapter 11, Nanomedicine, summarizes the application of nanotechnology in nanomedicine, such as areas of nanopharmaceuticals, nanoopthalmology, and regenerative medicine in addition to the bone tissue engineering.

AN INTRODUCTION TO NANOTECHNOLOGY*

1.1 DEFINITION

There is no fixed definition for nanotechnology that is precise and clear, because it is still a growing field with many areas that are still unknown. In 1999, the definition of nanotechnology set out from the National Nanotechnology Initiative (NNI) is given from an article (Roco, 1999):

> Nanotechnology is the ability to control and restructure the matter at the atomic and molecular levels in the range of approximately 1–100 nm, and exploiting the distinct properties and phenomena at that scale as compared to those associated with single atoms or molecules or bulk behaviour. The aim is to create materials, devices, and systems with fundamentally new properties and functions by engineering their small structure. This is the ultimate frontier to economically change materials properties, and the most efficient length scale for manufacturing and molecular medicine. The same principles and tools are applicable to different areas of relevance and may help establish a unifying platform for science, engineering, and technology at the nanoscale. The transition from single atoms or molecules behaviour to collective behaviour of atomic and molecular assemblies is encountered in nature, and nanotechnology exploits this natural threshold.

We can see that the definition is very long and detailed. Over a decade ago, there were not as many nanotechnology discoveries as there are today, and it can be shown by the change in definition. In 2010, the International Standardization Organization (ISO) Technical Committee 229 on nanotechnologies (ISO, 2010) issued a definition of nanotechnology:

> The application of scientific knowledge to manipulate and control matter in the nanoscale range to make use of size- and structure-dependent properties and phenomena distinct from those at smaller or larger scales.

The more recent definition is precise and clear. The elements chosen in the definition are the manipulation and control of matter on the nanoscale and the creation of materials and devices using the smaller scale. Basically, we can say that nanotechnology is the creation of functional materials, devices, and systems through control of matter on an atomic molecular scale.

---

*By Yaser Dahman, Hok Heng Lo, and Mark Edney.

Nanotechnology and Functional Materials for Engineers. DOI: http://dx.doi.org/10.1016/B978-0-323-51256-5.00001-0

1.2 **INTRODUCTION**

One of the most important reasons for studying nanotechnology is that it is on this nanoscale that the fundamental physical and chemical properties of a material and system are determined. These fundamental properties include its melting temperature, thermal conductivity, charge capacity, electronic conduction, tensile strength, and even color. This creates an interesting situation in which a material may have one set of properties on the large scale and a different set of properties on the nanoscale. The reason for this is that there is continuous modification of material characteristics with its changing size. The onset of nanotechnology illustrates how small our understanding is on the detailed processes by which molecules organize and assemble themselves. This leads into the exciting field of the construction of quantum devices and the operation of complex nanostructured systems. It is this change in properties based on material size that lead to exploration of a wide new array of different possibilities that were once never conceived, such as opaque substances that become transparent (copper), stable material that turns combustible (aluminum), solids that turn into liquids at room temperature (gold), and insulators that become conductors (silicon). One of the more intriguing discoveries is gold which is chemically stable on the large scale, but when examined on the nanoscale, it can serve as a very potent chemical catalyst. Much of the interest in nanotechnology is due to the fact that the characteristics at the nanoscale greatly differ from the larger scale with which we are familiar.

Another interesting characteristic of nanotechnology is that it is not limited solely to one small field of science. The implications of changing the physical and chemical properties of a material can influence many existing fields of science and technology (Fig. 1.1). It is nearly impossible to completely understand the potential of nanotechnology on the future of humanity. Here is a list of only a few of potential beneficial uses for nanotechnology (Fig. 1.2):

- Vastly improved delivery and control characteristics as medicines move down to the nanoscale.
- Greatly increased printing accuracy as nanoparticles are employed which have properties of both dyes and pigments.

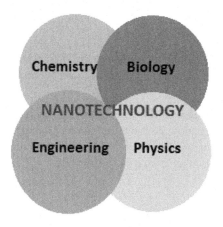

FIGURE 1.1

Nanotechnology involves different fields of science and engineering.

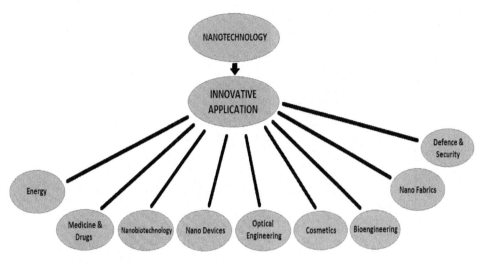

FIGURE 1.2

There is a variety of applications to nanotechnology.

- Vastly improved lasers and magnetic heads as increased control of layer thickness is moved toward the nanoscale.

There are two approaches for performing research within the field of nanotechnology: the top–down approach and the bottom–up approach. The top–down approach is characterized by a material being processed in bulk which is shaped into the finished product. In this approach, the positioning of individual atoms is not controlled during operation. Since no specific order in atoms can be achieved, this process results in defects and impurities (Forrest, 2008). Micropatterning techniques, such as photolithography and inkjet printing, belong to this category. Photolithography is a process of micro-fabrication employing the selective removal of a thin film from the surface of the sample. The bottom–up approach on the other hand describes a manufacturing process in which it is possible to control individual atoms. Bottom–up approaches rely on either (1) the chemical properties of a single molecule to self-organize or self-assembly in some useful configuration or (2) positional assembly. Self-assembly offers huge potential economic advantages. The creation of nanocomposites with organic molecules can be performed by depositing organic molecules to ultrasmall particles or ultrathin man-made layered structures (Arnall, 2003) (Fig. 1.3).

1.3 **HISTORY**

Although nanotechnology is a relatively recent development in scientific research, the development of the central concepts has happened over a longer period of time. We have unknowingly been using nano-technology for thousands of years through applications such as making steel, paintings, and in vulcan-ized rubbers. The development of the body of concepts that we understand to be called nanotechnology

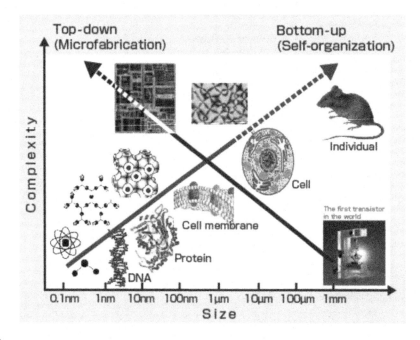

FIGURE 1.3

The types of fabrication of nanomaterials.

From AIST (2007). http://www.aist.go.jp/index_en.html.

however has only recently turned up. The first mention of some of the distinguishing concepts in nanotechnology appeared in 1867 by James Clerk Maxwell. He proposed a thorough experiment known as Maxwell's demon which is a tiny entity able to handle individual molecules. The first observations on the nanoscale took place in the first decade of the 20th century by Richard Adolf Zsigmondy. He was preforming a detailed study on gold sols and other nanomaterials with sizes 10 nm and less, which was published in 1914. He was able to determine particle sizes smaller than the wavelength of light through the use of ultramicroscopes which employed the dark field method. He was also the first person to use the nanometer for characterizing particle sizes which he determined to be 1/1,000,000 of a millimeter. He developed the first system of classification based on the particle size in the nanometer range. Another important historical event in the field of nanotechnology was from Irving Langmuir and Katharine B. Blodgett who first introduced the concept of the monolayer, a layer of material one molecule thick for which Irving won the noble prize.

The conceptual origins of nanotechnology can be found in a talk presented by the physicist Richard Feynman (pictured in Fig. 1.4) at the American Physical Society meeting at Caltech on December 29, 1959 titled "There's Plenty of Room at the Bottom" (Lewis, n.d.). In this talk, Richard introduces the idea of a process by which the ability to manipulate individual atoms and molecules might be developed using one set of precise tools to build and operate another proportionally smaller set, and so on down to the needed scale. He also had the foresight to realize some of the many different phenomena that come with the nanoscale, such as gravity's less important role and surface tension's and Van der

FIGURE 1.4

Richard Feynman.

From Nobel Media AB (2014). Richard P. Feynman—Biographical. Retrieved 12.05.16, from
https://en.wikipedia.org/wiki/File:Richard_Feynman_Nobel.jpg

Waal attraction forces' more important one. At the end of his talk, he proposed two challenges for which he offered a cash price: one was the construction of a nanomotor which was achieved in 1960 by William McLellan and the second involved the possibility of scaling down letters small enough to be able to print the entire Encyclopedia Britannica on the head of a pin. The prize for the second challenge was accepted by Tom Newman in 1985.

The next important historical moment took place in 1965 when Gordon Moore observed that silicon transistors were undergoing continual scaling down, which was later codified as Moore's Law. Since his observation in 1965, transistor sizes have decreased from $10 \mu m$ to the range of 45–60 nm as of 2007.

The term "Nanotechnology" was defined much later in a 1974 paper by Norio Taniguchi of the Tokyo Science University (Lewis, n.d.). In this paper he stated "Nano-technology mainly consists of the processing of separation, consolidation, and deformation of materials by one atom or one molecule." Since that time however, the definition of nanotechnology has become more general and includes features as large as 100 nm. The idea that nanotechnology embraces structures exhibiting quantum mechanical aspects has further evolved it definition. Also in the year 1974, Dr. Tuomo Suntola developed and patented the process of atomic layer deposition used for depositing uniform thin films of one atomic layer at a time.

Dr. K. Eric Drexler in 1980 explored the idea that nanotechnology was deterministic rather than stochastic, meaning that nanotechnology is not based on random elements. He promoted the technological significance of nanoscale phenomena and devices through many of his speeches and books. Drexler's vision of nanotechnology is often called "molecular nanotechnology" or "molecular manufacturing."

Richard Jones in 2004 wrote "Soft Machines" in which he describes radical nanotechnology as a deterministic idea of nanoengineering machines that do not take into account nanoscale challenges such as wetness, stickiness, Brownian motion, and high viscosity. He explains that soft nanotechnology or biomimetic nanotechnology is the best way to design functional nanodevices that can cope with all the problems at the nanoscale. Soft nanotechnology can be the application of learning lessons from biology on how things work, chemistry to precisely engineer such devices, and stochastic physics to model the system and its natural processes in detail.

There have been many important advances in different experiential equipment that have played a drastic role in the changing field of nanotechnology. Two of the very first important technological advances where the birth of cluster science and the invention of the scanning tunneling microscope (STM) which took place in the early 1980s. STM is an instrument of imaging surfaces at the atomic level invented by Gerd Binnig and Heinrich Rohrer for which they won the Nobel Prize. A good resolution for an STM is considered 0.1 nm lateral resolution and 0.01 nm depth resolution. With this level of resolution, individual atoms within materials are routinely imaged and manipulated. The development of the STM has led to many discoveries such as fullerenes in 1985 and a few years later, nanotubes. Fullerenes are molecules that are composed entirely of carbon and form a hollow sphere. Nanotubes are molecules of carbon that are in the form of cylinders. Through the studying and synthesis of nanocrystals with an STM, there were great advances in the field of semiconductors. Don Eigler was the first person that to manipulate atoms using STM in 1989. He used 35 Xenon atoms to spell out the IBM logo (Fig. 1.5).

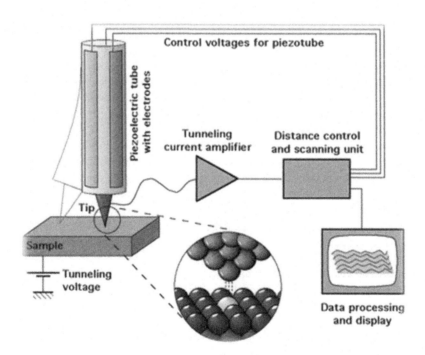

FIGURE 1.5

Schematic of an STM.

From Wikipedia. (n.d.). Wikipedia. Retrieved 13.04.11, from Scanning tunneling microscope: http://en.wikipedia.org/wiki/Scanning_tunneling_microscope.

It was in the early 1990s when Huffman and Kratschmer from the University of Arizona first discovered how to synthesize and purify large quantities of fullerenes. This had opened the door to a wide area of investigation into the characterizing and functionalizing of fullerenes. Shortly after this, it was found that C_{60} can become a mid-temperature superconductor when doped with rubidium. It was a speech presented in 1992 by Dr. T. Ebbesen in which he spellbound his audience of his discovery and characterization of carbon nanotubes (CNTs), stimulating many to go to their laboratories to reproduce and push his discoveries. It was this speech that spurred the interest in hundreds of researchers that helped further develop the field of nanotube-based nanotechnology. At present the practice of nanotechnology embraces both stochastic approaches and deterministic approaches wherein single molecules are manipulated on the substrate surface through the use of STM or an atomic force microscope (AFM). The AFM is based on a cantilever with a sharp tip that is used to scan the surface of the specimen. The cantilever itself is typically silicon or silicon nitride with a tip radius of curvature on the nanoscale. When the surface of the cantilever is brought in proximity to the sample surface, forces between the tip and the surface sample lead to deflection of the cantilever according to Hooke's law (Fig. 1.6).

Another important historical event within the field of nanotechnology is the NNI that was passed in 1999. The initiative had four goals, which are the following:

1. Advance a world-class nanotechnology research and development program.
2. Foster the transfer of new technologies into products for commercial and public benefit.
3. Develop and sustain educational resources to advance nanotechnology.
4. Support responsible development of nanotechnology.

This initiative has made the United States a lead in the field of nanotechnology.

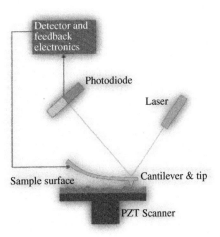

FIGURE 1.6

Atomic force microscope.

From Wikipedia (2011).

1.4 APPLICATIONS

The research trends can be understood by taking a look at publication trends. The number of papers reflecting discoveries in the area of nanotechnology has reached about 65,000 in 2008, compared to 18,000 in 2000. The rapid increase shows the growing interest in nanotechnology.

New processes and nanostructures have been created using principles from quantum and surface sciences to molecular bottom–up assemblies. They are combined with top–down techniques integrated into products. Some examples of using nanotechnology in the areas of computing, quantum computing, and communication devices can be seen in semiconductors, thin film storage (TFS), and magnetic random access memory.

Metal oxide semiconductor processor technology by Intel is used to make integrated circuit chips that are available in a wide variety of laptop, desktop, and server computer systems, giving higher speed, higher density, and lower power consumption using quantum dot mechanics.

TFS flash memory by Freescale is used in microcontrollers, utilizing silicon nanocrystals as the charge storage layer. The nanocrystal layer enables higher density arrays, lower power operation, faster erase times, and improved reliability. Microcontrollers are the brains of a variety of industrial and consumer products.

Magnetic random access memory from Everspin is based on nanometer-scale magnetic tunnel junctions. These memories have many industrial applications such as saving data during a system crash, enabling resume-play features, quick storage, and retention of data encryption during shutdown.

Nanoscale medicine has made significant breakthroughs and is advancing rapidly. There are applications of biocompatible materials, diagnostics, and treatments. Abraxane is commercializing advanced therapeutic treatments for different forms of cancer. There are over 50 cancer-targeting drugs based on nanotechnology that are in clinical trial in the United States alone.

Fig. 1.7 shows examples of commercial healthcare products that are FDA approved and have nanotechnology incorporated into the treatment or product. Beginning clockwise at the top, the Nanosphere Verigene system is used for onsite medical diagnostics that use gold nanoparticle technology to detect nucleic acid and protein targets of interest for a wide variety of applications. Luna nanoparticle contrast agents give enhanced clarity and safety of diagnostic magnetic resonance imaging using nanomagnetic tracing. NanOss from Angstrom Medica uses nanocrystalline calcium phosphate as synthetic bone material for replacements/reinforcements. Provenge from Dendreon is an immunotherapy product made using cells from a patient's own immune system to fight prostate cancer. Abraxane from Celgene uses nanoparticle albumin bound technology to leverage albumin nanoparticles for the active and target delivery of chemotherapeutics to treat metastatic breast cancer.

There has been extensive penetration of nanotechnology into many industries, including catalysis by engineered nanostructured materials. In the oil and chemical industries, 30–40% of the market includes products/catalysts created involving nanotechnology.

Fig. 1.8 shows the redesigned mesoporous silica materials, like MCM-41, along with zeolites, which are used in a variety of process such as fluid catalytic cracking for producing gasoline from heavy gas oils, and transalkylation for producing para-xylene and related building blocks for the manufacturing of polyesters.

Research trends have shown key themes in the integration of nanotechnology in consumer and industrial products.

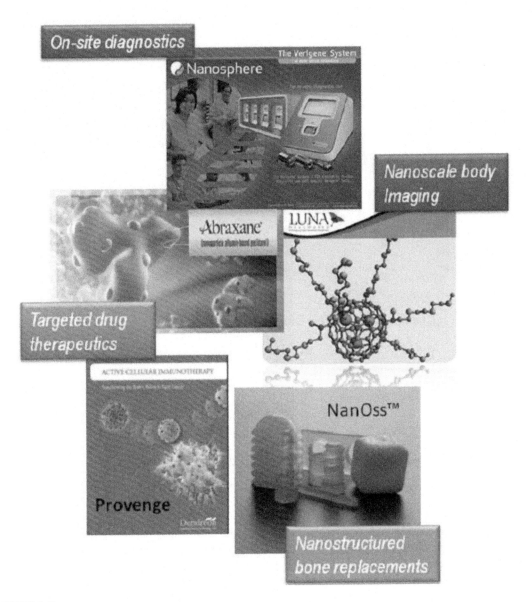

FIGURE 1.7

Healthcare products that incorporate nanotechnology for the diagnosis and treatment of certain medical conditions and diseases.

From Roco, M. (2011). Nanotechnology research directions for societal needs in 2020: summary of international study. Journal of Nanoparticle Research.

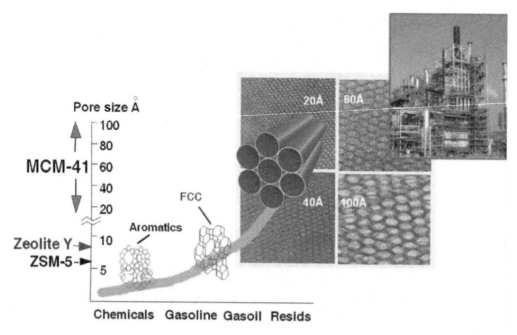

FIGURE 1.8

Redesigned mesoporous silica materials.

From Roco, M. (2011). Nanotechnology research directions for societal needs in 2020: summary of international study. Journal of Nanoparticle Research.

1.5 RECENT RESEARCH AND FUTURE PERSPECTIVES

There is no doubt that nanotechnology is an emerging field with much to come for many years. As discussed throughout this chapter, the number of published papers has increased exponentially in the last few years. Kohler, Mietke, Ilgner, and Werner (2003) discuss nanotechnology with respect to the markets and trends regarding nanomaterials and CNTs. They state that nanotechnology is a very general term and that there is no generally accepted definition. That makes it challenging to predict the world market for products produced by nanotechnology. They discuss nanotechnology as all products with a controlled geometry size of at least one functional component below 100 nm in one or more dimensions that makes physical, chemical, or biological effects available. Market forecasts from their article show that the investment trend is increasing exponentially for nanomaterials. Fig. 1.9A shows the possible future scenario for the time-to-market of various applications. Fig. 1.9B shows the market for nanomaterials from 2002 to 2006.

The authors claim that nanotechnology played an important role in different application fields in an evolutionary way. At the time of publication, nanotechnology was penetrating a number of products. Nanomaterials and nanotools were already in the market and had the best opportunity to be commercialized. The market and product forecasts were overestimated from a long-term perspective; it was

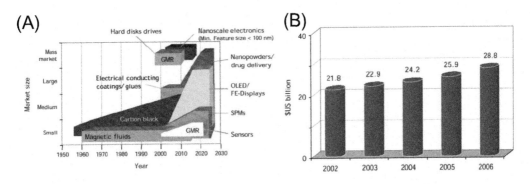

FIGURE 1.9

(A) Possible future scenario for the time-to-market of selected applications. (B) Nanomaterials market forecast in US dollars from 2002 to 2006.

From Kohler, T., Mietke, S., Ilgner, J., & Werner, M. (2003). Nanotechnology—markets & trends.

Table 1.1 Effects and Applications Due to Reduced Dimension	
Effect	**Applications**
Higher surface to volume ratio, enhanced reactivity	Catalysis, solar cells, batteries
Increased hardness with decreasing grain size	Hard coatings, thin protection layers
Lower percolation threshold	Conductivity of materials
Narrower bandgap with decreasing grain size	Optoelectronics
Higher resistivity with decreasing grain size	Electronics
Increased wear resistance	Hard coatings, tools
Lower melting and sintering temperature	Processing of materials, low sintering materials
Improved transport kinetics	Batteries, hydrogen storage
From Kohler, T., Mietke, S., Ilgner, J., & Werner, M. (2003). Nanotechnology—markets & trends.	

believed that nanoelectronics based on novel concepts, such as CNTs, were in good position to become a market leader. There are special effects and applications when dimensions are reduced, as shown in Table 1.1.

The effects of the reduced dimensions compared with the applications show the possibilities of discovery in nanotechnology in years following the publication of the article. The authors also explain that single-walled CNTs properties are much better than the non-CNT infused material. From their perspective, CNTs were in a good position to become a market leader in nanoelectronic-based concepts and materials.

The exploit of the full potential of nanotechnology requires a significant amount of research required in the areas of nanotechnology. Nanotechnology is referred to as a combination of nanomaterials science, physics, and biotechnology into a compact entity. Kohler et al. urge a close collaboration between biologists, chemists, physical scientists, and engineers to push further advances in bio- and nanotechnology.

Table 1.2 Six Key Indicators of Nanotechnology Development in the World and the United States

World (USA)	People Primary Workforce	SCI Papers	Patent Applications	Final Products Market	R&D Funding Public + Private	Venture Capital
2000 (actual)	60,000 (25,000)	18,085 (5342)	1197 (405)	$30 B ($13 B)	$1.2 B ($0.37 B)	$0.21 B ($0.17 B)
2008 (actual)	400,000 (150,000)	65,000 (15,000)	12,776 (3729)	$200 B ($80 B)	$15 B ($3.7 B)	$1.4 B ($1.17 B)
2000–2008 (average growth)	25%	23%	35%	25%	35%	30%
2015 (2000 estimate)	2,000,000 (800,000)			$1000 B ($400 B)		
2020 (extrapolation)	6,000,000 (2,000,000)			$3000 B ($1000 B)		

From Roco, M. (2010). The long view of nanotechnology development: the National Nanotechnology Initiative at 10 years.

The NNI of the United States of America, the `central point of communication, cooperation, and collaboration for all Federal agencies engaged in nanotechnology research had its 10 anniversary in the year 2010. Roco (2010) provided a perspective on the development of the NNI since the year 2000 in the international context, the main outcomes of the R&D programs after 10 years, the governance aspects specific to nanotechnology, lessons learned, and how the nanotechnology community should prepare for the future. In 2010, the ISO issued a definition of nanotechnology: The application of scientific knowledge to manipulate and control matter in the nanoscale range to make use of size- and structure-dependent properties and phenomena distinct from those at smaller or larger scales. Table 1.1 describes the six key indicators of nanotechnology development globally. These indicators help portray the value of investments in nanotechnology development and associated science breakthroughs and technological applications.

We can see from Table 1.2 that the columns represent the growth of nanotechnology with respect to knowledge, research, development, and investment. The number of researchers and workers involved in one domain or another of nanotechnology was estimated at about 400,000 in 2008. The estimate at about 2,000,000 in 2015 can be achieved if the growth rate of 25% continues. The number of Science Citation Index (SCI) papers reflecting discoveries in the area of nanotechnology reached about 65,000 in 2008 as compared to 18,000 in 2000, showing the rapid increase of the popularity and interest in nanotechnology. The inventions shown by the number of patent applications filed are about 13,000 in 2008, which is an annual growth rate of about 35% when compared to the 1200 patent applications in 2000. This shows that the discoveries in the nanoscale science and engineering aspects of nanotechnology are being explored and the knowledge base is growing quickly. The value of products incorporating nanotechnology as key components reached about $200 billion in value worldwide in 2008. This shows that the value of discoveries in nanotechnology is potentially limitless. Global nanotechnology research and development from both private and public sectors reached about $15 billion and continues to grow as much more interest is being shown in nanotechnology. The global venture capital investment

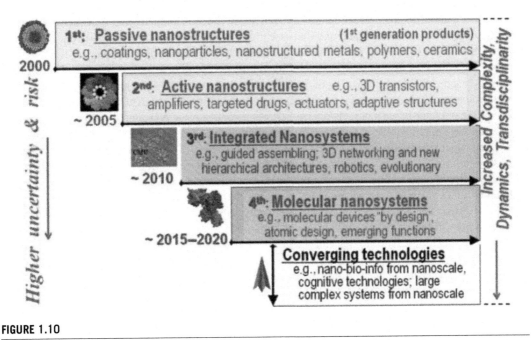

FIGURE 1.10

Timeline for the prototyping of nanotechnology.

From Roco, M. (2010). The long view of nanotechnology development: the National Nanotechnology Initiative at 10 years.

reached about $1.4 billion in 2008, where the majority was in the United States. From these growth figures, we can clearly see that interest in nanotechnology is growing with discoveries and knowledge being spread as they are being discovered.

With all the discoveries in nanotechnology, there is growing focus on the integration of the discoveries with current applications. Fig. 1.10 is a growth timeline of the stages of nanotechnology discovery and development.

There are objectives discussed that have not been fully realized by 2010, and others that have done better than expected. Sustainable development projects such as nanotechnology for energy solutions received momentum after 5 years and nanotechnology for water filtration has limited funding. On the bright side, the formation and growing strength of an international nanotechnology community, including in nanotechnology Environment, Health and Safety, as well as Ethical, Legal, and Societal Issues. These developments have surpassed expectations, and the debut of governance studies was unanticipated. The paper stresses that it will be imperative over the next decade to focus on four distinct aspects of nanotechnology development that are discussed in this volume: (1) better comprehension of nature leading to knowledge progress, (2) economic and social solutions leading to material progress, (3) international collaboration on sustainable development leading to global progress, and (4) people working together for equitable governance leading to moral progress. Roco (2010) has previously conducted studies on the development of nanotechnologies and continue to research on the trends and applications of nanotechnology.

The use of nanotechnology for medicine has been one of the most funded areas of research. Murday et al. (2009) gave a status assessment and discussed future opportunities for nanomedicine. The 2007 cost of the medical system in the United States consists of $90 billion being spent on cancer treatment, $116 billion on diabetes treatment, and $1.5 billion on the treatment of spinal cord injury. The amount of implants reached 500,000, consisting of 200,000 hip implants and 300,000 knee implants.

Nanotechnology has made a great emergence in the field of imaging with contrast agents including biodegradable polymer-based nanogels/nanospheres/nanoemulsions, CNTs, dendrimers, gold nanoparticles, liposomes, micro bubbles, semiconductor quantum dots, silica nanoparticles with enclosed fluorescers, and super paramagnetic iron oxide particles. The purpose of these innovations are to enhance reduced signal-to-noise sensitivity and resolution levels, for selective binding to target cells, for acceptable toxicity profile, and for their ease of production.

Another toll in diagnostics that is provided by nanotechnology is the use of microelectromechanical systems (MEMS). These MEMS can convert medically relevant parameters of biological events into electrical signals. Nanotechnology has also revolutionized in vitro diagnostics with microarrays, which are miniaturized chip-based arrays that can detect at higher sensitivity than traditional laboratory equipment. Laboratory-on-a-chip will require only a small volume of sample and reagents, thus producing less waste and decreasing analysis times, while remaining relatively cheap with reduced dimensions. Drug delivery has also been improved with greater control on drug release with the result of reducing negative side effects.

Only 1 of every 100,000 molecules of agents reaches its desired destination through the current intravenous drug delivery method, so there is much room for improvement. Implants and tissue regeneration can be improved with the introduction of engineered scaffolds, which help the body grow on its own. Even dental restorative techniques have been improved with nanotechnology by incorporating nanoparticles.

One of the first proposed applications for nanotechnology within the field of biomedicines is the idea of DNA-based noninvasive vaccinations. Non-invasive vaccination on the skin (NIVS) would not require specially trained personnel and would eliminate many problems associated with needle injections. Shi et al. explain that the delivery by pipetting adenovirus or liposome-complexed plasmid DNA onto the outer layer of the skin could achieve localized expression. There has already been proven success of NIVS with protein-based vaccines. DNA-based vaccines can be purified at lower costs and they may be able to stimulate a broader spectrum of immune system responses. During their experiment, the NIVS delivery system was found to work, but it was not as effective as the traditional IM method. There are still some potential benefits to this delivery method as it is simple, economical, painless, and potentially very safe.

Nanodelivery systems, in general, are heavily studied and researched. Maeda Mamiya et al. (2009) introduced a new DNA-delivering system. Fullerene is a compound based on carbon atoms that are connected together in a spherical shape. Gene delivery can be achieved by chemically modifying the fullerene structure into tetra (piperazino) fullerene epoxide (TPFE). This compound has many advantages when compared to the present gene delivery system, lipofection. Lipofection is based on the use of liposomes that act as a sheath around the DNA connecting them by ionic interactions. The unique features of the fullerene are its spherical structure and high hydrophobicity, making it a desirable candidate. The method introduced by Mamiya et al. experienced better absorption within the kidneys, liver, and spleen but a decrease in the lungs when compared to Lipofection. Liver injury is measured by higher than normal concentrations of aspartate aminotransferase (AST) and alanine aminotransferase (ALT). Both AST and ALT were found in the samples within the Lipofection group, but there were

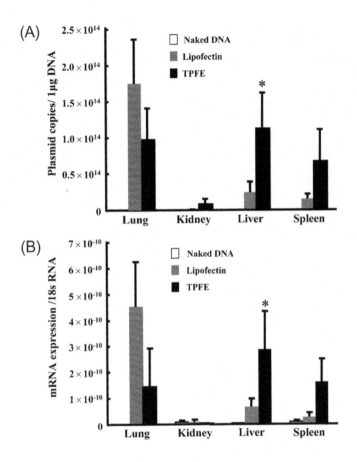

FIGURE 1.11

Distribution of injected plasmid DNA and enhanced green fluorescent protein (EGFP) mRNA expression in each organ. (A) Plasmid copies and (B) mRNA expression. From both images, it can be seen that the results are favorable for TPFE in all organs, except the lung.

From Maeda-Mamiya, R., Noiri, E., Isobe, H., Nakanishi, W., Okamoto, K., Doi, K., et al. (2009).
In vivo gene delivery by cationic tetraamino fullerene.

none in the TPFE group. The lipofection group also experienced an increase in blood urea nitrogen levels potentially indicating acute kidney injury. Some results are shown in Fig. 1.11. Overall, the TPFE method did not appear to have a toxicity effect. It was also found that TPFE can be used for insulin gene delivery which can reduce glucose levels.

Although nanotechnology is new with a variety of properties and phenomena with which we are still unfamiliar, it is important to understand that some of the principles that we know in chemistry still apply. Just because we have recently added the word "nano" to some previous sciences or substances, it does not make them any different. Nowak, Krug, and Height (2011) state that many discussions are centered on the asserted assumption that nanoparticles are something fundamentally new and thus

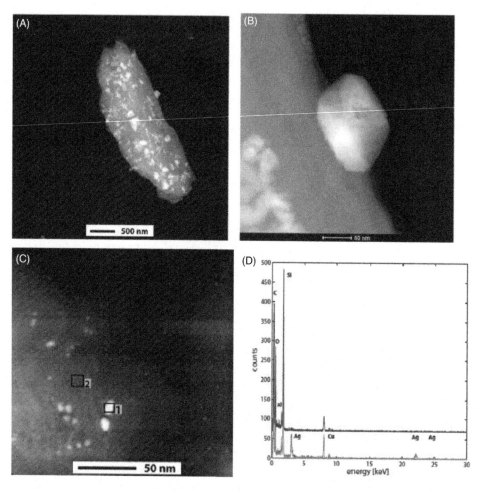

FIGURE 1.12

TEM analysis of silver-impregnated carbon filter. (A) larger particle with silver particles; (B) magnification of one silver nanoparticle from the picture to the left; (C) small silver nanoparticles on gray matrix; (D) EDX spectra of the two areas in the TEM picture to the left (bottom spectrum: silver particles, top spectrum: background).

From Nowak, B., Krug, H., & Height, M. (2011). 120 years of nanosilver history: implications for policy.

cannot be compared to conventional chemicals or bulk materials. Nanosilver is one of the nanomaterials that is under scrutiny today, and its release and effects are studied widely. However, EPA has registered (nanosilver particle-containing) silver-impregnated water filters since the 1970s! To verify the presence of nanoscale silver in water filters, commercially available filters were disassembled and the carbon pellets were crushed and investigated with a transmission electron microscopy (TEM). Fig. 1.12 shows an analysis of nanosilver particles from a 2002 US patent from Zodiac Pool Care Inc.

The analysis shows that bright nanoparticles that have sizes ranging from a few nanometers to 100 nm can be found in the tested sample. This analysis proves that during manufacturing of the silver-impregnated water filters, silver nanoparticles are formed and are present in the filters. These filters have been safely used for domestic water and no health or environmental effects have been reported. Nowack et al. have shown and argued that nanosilver has not been harmful in past uses and thus regulation should be reassessed.

1.6 CONCLUSIONS

Given the development of this emerging field, we can see that there is an enormous amount of discoveries possible in nanotechnology. Research trends have shown that consumer and industrial products are being integrated with nanotechnology more and more which shows the value of the smaller scale technology.

Nanotechnology is still in its early phase of development and fundamental understanding and tools are still in the stream of new ideas and innovations. Key research themes have helped nanotechnology develop as much as it has over the last decade. To summarize, the future outlook of nanotechnology trends shows the stages of nanotechnology discovery and integration of nanosystems and the converging of technologies. The converging technologies increase the risk, complexity, and dynamics of nanotechnology.

It will be vital over the next decade to focus on four distinct aspects of progress in nanotechnology: (1) How nanoscale science and engineering can improve understanding of nature, protect life, generate breakthrough discoveries and innovation, predict matter behavior, and build materials and systems by nanoscale design—knowledge progress; (2) How nanotechnology can generate medical and economic value—material progress; (3) How nanotechnology can promote safety in society, sustainable development, and international collaboration—global progress; and (4) How responsible governance of nanotechnology can enhance quality of life and social equity—moral progress.

GENERIC METHODOLOGIES FOR CHARACTERIZATION*

2.1 INTRODUCTION

Nanotechnology is the term used to cover the design, construction, and utilization of functional structures with at least one characteristic dimension measured in nanometers (Kelsall, Hamley, & Geoghegan, 2005). Therefore, in order to study these nanoparticles, generic methodologies for characterization have been developed or improved since the beginning of the nanotechnology era. These characterization techniques (Kelsall et al., 2005) are summarized in Fig. 2.1.

Before using a particular characterization technique, it is essential to consider what information is required about a sample. This information can be divided into (1) morphology (microstructural or nanostructural architecture), (2) crystal structure (detailed atomic arrangement), (3) chemistry (elements and molecular groups present), and (4) electronic structure (the nature of the bonding between atoms). In summary, these techniques can be used for imaging (microscopy) and analysis (spectroscopy) of samples (Kelsall et al., 2005).

In this chapter, we will introduce theoretical background about some of these techniques (the ones in red (grey in print version) given in Fig. 2.1) as well as their uses for nanotechnology research purposes. This chapter is outlined differently than most others because all of the background theory, applications, and the future perspectives are listed for each method of characterization. Table 2.1 presents a summary of the characterization techniques available and their typical uses.

2.2 SCANNING PROBES

Scanning probe microscopy (SPM) is a family of mechanical probe microscopes that measures surface morphology in real space with a resolution down to the atomic level. SPM was originated from scanning tunneling microscopy (STM), in which the electrical current caused by the tunneling of electrons through the tip and the sample was used as the feedback parameter to maintain separation between them. This STM technique, invented in 1981, was a totally new one since it could image atom arrangement on a surface in real space for the first time. It is so invaluable to science and technology that the inventors of STM shared the Nobel Prize in Physics with the inventor of the electron microscopy in 1986 (Heng-Yong, 2010).

*By Yaser Dahman, Caroline Halim, Oswaldo Matos, and Louisa Chan.

Nanotechnology and Functional Materials for Engineers. DOI: http://dx.doi.org/10.1016/B978-0-323-51256-5.00002-2

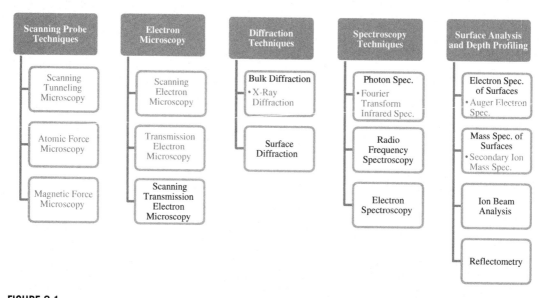

FIGURE 2.1

Generic methodologies for nanotechnology characterization.

2.2.1 SCANNING TUNNELING MICROSCOPY

The principle of the STM may be simple: tunneling of electrons between two electrodes under an electric field. However, to develop the concept of electron tunneling into a technology of imaging atomic resolution on a surface was not simple. To measure the tunneling current, the distance between the two electrodes must be close to each other on the order of 1 nm. Surface cleanness and a vibration-free system are essential to measuring the tunneling current accurately (Heng-Yong, 2010).

Fig. 2.2 shows a diagram depicting how STM functions. By approaching the tip with a specified bias and current, the tip will be held at a certain distance from the sample surface so that the specified current (set point) is realized. By scanning the tip across the sample under this condition, the system compares the measured current I and the set point current I_s. It uses the error signal $I-I_s$ as the feedback parameter to apply an appropriate voltage to the z-piezo to adjust the tip–sample distance so as to diminish the error signal (i.e., $I-I_s \sim 0$), thus providing the height profile of the "topography" of the surface. This is the constant current mode. The other operation (constant height) mode is to keep the tip–sample distance while recording the current, which apparently requires the scanned area to be flat (Heng-Yong, 2010).

Xu, Cao, and Heath (2009) presented an STM study of a new class of corrugations in monolayer graphene sheets that have been largely neglected in previous studies, that is, wrinkles approximately 10 nm in width and approximately 3 nm in height. Through STM, they also demonstrated that these wrinkles had lower electrical conductance when compared to other regions of graphene, and therefore, they likely play an important role in determining the electrical properties of graphene sheets. The approximately 20 μm monolayer graphene sheets (investigated in this journal) were fabricated on insulating SiO_2 substrates through mechanical exfoliation of Kish graphite flakes.

Table 2.1 Summary of the Characterization Techniques Most Commonly Used for the Analysis of Nanostructures and Nanomaterials

Characterization Technique	Primary Agent	Secondary Effect Excited or Signal Detected	Typical Applications
Electron microscopies			
TEM, HRTEM, STEM	Electrons	Transmitted electrons, X-rays	Imaging, elemental composition (in combination with EDS), study of crystal structure
SEM, HRSEM	Electrons	Backscattered and secondary electrons, X-rays	Imagining, elemental composition (in combination with EDS/EDX)
LEEM	Electrons	Scattered electrons	Determination of crystal structure
Other electron techniques			
RHEED	High-energy electrons	Scattered electrons	Determination of crystal structure
LEED	Low-energy electrons	Scattered electrons	Determination of crystal structure
EELS, LEELS	Electrons	Transmitted electrons	Elemental mapping (generally in combination with TEM, HRTEM, or STEM)
Spectroscopic techniques			
Raman spectroscopy, SERS	Photons	Scattered photons	Chemical composition
XPS (UPS)	Low energy X-rays (ultraviolet photons)	Photoelectrons	Elemental identification and quantification, depth profiling
AES	Electrons	Auger electrons	Elemental identification and quantification, depth profiling
Scanning probe microscopies			
NSOM	Photons	Photons	Topography, optical properties
STM	A conducting tip scans the surface	Tunneling current	Topography (conducting materials
AFM	A tip scans the surface	Atomic forces	Topography
Magnetic resonance techniques			
NMR	Magnetic fields	Spins of atomic nuclei	Chemical analysis
EPR (ESR)	Magnetic fields	Electron spins	Chemical analysis
Ion-based techniques			
RBS	High-energy ions	Backscattered ions	Elemental composition depth profiling
PIXE (PIGE)	Ions	X-rays (gamma rays)	Elemental composition, mapping (microPIXE)

(Continued)

Table 2.1 Summary of the Characterization Techniques Most Commonly Used for the Analysis of Nanostructures and Nanomaterials (Continued)

Characterization Technique	Primary Agent	Secondary Effect Excited or Signal Detected	Typical Applications
ERDA	Ions	Recoiled atoms	Elemental composition, depth profiling (light elements)
SIMS	Ions	Secondary ions	Elemental isotopic composition, depth profiling
NRA	Ions	Reaction products	Elemental composition, depth profiling
Other techniques			
XRD	X-rays	Diffracted X-rays	Crystal structure, nanocrystal size, orientation distribution
Infrared spectroscopy (IR, FTIR)	IR photons	IR photons	Composition
Optical spectrometry	Photons	Reflected and transmitted photons	Optical properties
Ellipsometry	Photons	Reflected photons	Optical properties
Luminescence, fluorescence	Photons	Emitted photons	Composition

From Martin-Palma, R. J., & Lakhtakia, A. (2010). Nanotechnology: A crash course. Washington: SPIE.
TEM, transmission electron microscopy; HRTEM, high-resolution transmission electron microscopy; EDS/EDX, energy-dispersive X-ray spectroscopy

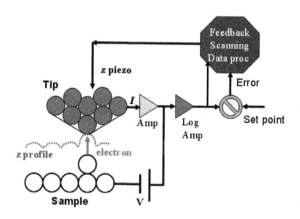

FIGURE 2.2

Schematic diagram of STM.

From Heng-Yong, N. (2010, November 5). Scanning probe microscopy.

On wrinkle   "Flat" part

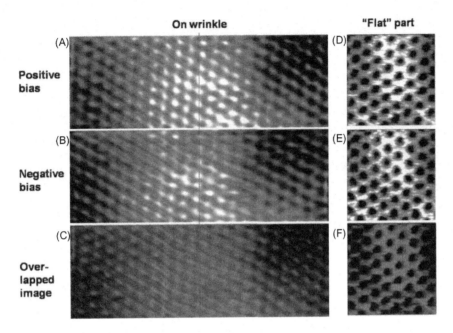

FIGURE 2.3

Comparison of STM topographs of a graphene wrinkle and a "flat" part of the same graphene sheet.

From Xu, K., Cao, P., & Heath, J. R. (2009). Scanning tunneling microscopy characterization of the electrical properties of wrinkles in exfoliated graphene monolayers. Nano Letters.

Fig. 2.3A and B gives the STM topographs of the top surface of a graphene wrinkle, obtained at positive and negative sample biases. The same "three-for-six" triangular patterns were observed for both sample biases. The center parts of the two topographs (crest of the wrinkle) overlapped with each other exactly, and a regular triangular lattice was obtained (see Fig. 2.3C). In comparison, on the "flat" part of the same graphene sheet (Fig. 2.3D–F), although approximately 4 Å ripples were present, clear honeycomb structures were observed for both positive and negative biases, and since these two also overlap exactly, this shows that the ripples are more rigid compared to the wrinkle.

The authors also found distinctly different electrical properties for the approximately 3 nm high wrinkles (Fig. 2.4A). Fig. 2.4B shows the differential conductance behavior of a wrinkle, in comparison with other parts of the graphene sheet, where small ripples were present. Lower conductance was observed for the wrinkle at low bias voltage, indicating the wrinkle was less conductive than the other parts of graphene. This observation agreed with the theory that local bending/curvature effects weaken delocalized π-bonds.

2.2.2 ATOMIC FORCE MICROSCOPY

Microscopes have always been one of the essential instruments for research in the biomedical field. Radiation-based microscopes (such as the light microscope and the electron microscope) have become

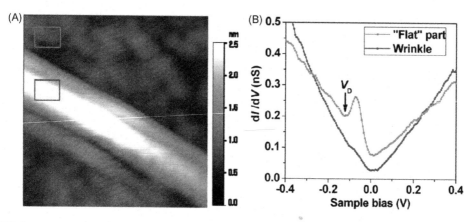

FIGURE 2.4

(A) STM topograph (50 nm × 50 nm) of the wrinkle and (B) differential conductance of the wrinkle.

From Xu, K., Cao, P., & Heath, J. R. (2009). Scanning tunneling microscopy characterization of the electrical properties of wrinkles in exfoliated graphene monolayers. Nano Letters.

trustworthy companions in the laboratory and have contributed greatly to our scientific knowledge. However, although digital techniques in recent years have still enhanced their performance, the limits of their inherent capabilities have been progressively reached. The advent of scanning probe microscopes and especially of atomic force microscopes has opened new perspectives in the investigation of biomedical specimens and induces to look again with rejuvenated excitement at what we can learn by "looking" at our samples. Novices are at first mesmerized by two features: the name of the instrument and the colorful 3D computer visualization of surfaces. One later learns that quite often it is not possible to obtain the "atomic" resolution that one hoped to achieve but that nevertheless images do contain some details not observed with any other instrument. The 3D mapping of the surface gains scientific relevance when one realizes that it is not just fancy surface reconstruction but that true topographic data with vertical resolution down to the nanometer range is readily available (Braga & Ricci, 2003).

2.2.2.1 AFM instrumentation

2.2.2.1.1 The microscope

Fig. 2.5 shows a schematic diagram of an AFM. In principle, AFM can bring to mind the record player, but it incorporates a number of refinements that enable it to achieve atomic-scale resolution, such as very sharp tips, flexible cantilevers, a sensitive deflection sensor, and high-resolution tip sample positioning.

2.2.2.1.2 The tip and cantilever

The tip, which is mounted at the end of a small cantilever, is the heart of the instrument because it is brought in closest contact with the sample and gives rise to the image though its force interactions with the surface. When the first AFM was made, a very small diamond fragment was carefully glued to one end of a tiny piece of gold foil (Braga & Ricci, 2003).

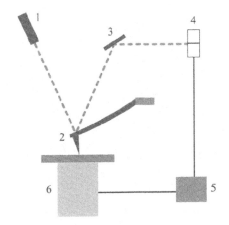

FIGURE 2.5

Schematic diagram of AFM.

From Braga, P. C., & Ricci, D. (Eds.), (2003). Atomic force microscopy: Biomedical methods and applications. Totowa, NJ: Humana Press. ProQuest ebrary. Web. 9 May 2016.

Today, the tip–cantilever assembly typically is fabricated from silicon or silicon nitride, and using technology similar to that applied to integrated circuit fabrication, allows a good uniformity of characteristics and reproducibility of results. The essential parameters are the sharpness of the apex, measured by the radius of curvature, and the aspect ratio of the whole tip. Although it would seem that sharper tips should yield more detailed images, this may not occur with all samples. In fact, quite often, so-called "atomic resolution" on crystals is obtained best with standard silicon nitride tips. In general, one can choose among one of three types of tip. The standard tip is usually a 3 μm tall pyramid with an approximately 30 nm end radius. Tips made from silicon (either polysilicon or single crystal) through improved microlithographic techniques have a higher aspect ratio and small apex radius of curvature, maintaining reproducibility and durability. The cantilever carrying the tip is attached to a small glass "chip" that allows easy handling and positioning in the instrument. There are essentially two designs for cantilevers, the "V" shaped and the single-arm kind (Fig. 2.6). They have different torsional properties. The length, width, and thickness of the beam(s) determine the mechanical properties of the cantilever and have to be chosen depending on the mode of operation needed and on the sample to be investigated. Cantilevers are essentially classified by their force (or spring) constant and resonance frequency: soft and low-resonance frequency cantilevers are more suitable for imaging in contact and resonance mode in liquid, whereas stiff and high-resonance frequency cantilevers are more appropriate for resonance mode in air (Braga & Ricci, 2003).

2.2.2.1.3 Deflection sensor

AFMs can generally measure the vertical deflection of the cantilever with picometer resolution. To achieve this, most AFMs today use the optical lever or beam-bounce method, a device that achieves resolution comparable to an interferometer while remaining inexpensive and easy to use. In this system, a laser beam is reflected from the backside of the cantilever (often coated by a thin metal layer to

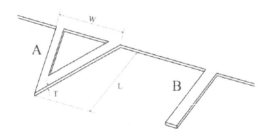

FIGURE 2.6

Triangular (A) and single-beam (B) cantilevers.

From Braga, P. C., & Ricci, D. (Eds.), (2003). Atomic force microscopy: Biomedical methods and applications. Totowa, NJ: Humana Press. ProQuest ebrary. Web. 9 May 2016.

make a mirror) onto a position-sensitive photodetector consisting of two side-by-side photodiodes. In this arrangement, a small deflection of the cantilever will tilt the reflected beam and change the position of the beam on the photodetector. The difference between the two photodiode signals indicates the position of the laser spot on the detector and thus the angular deflection of the cantilever. Because the distance between cantilever and detector is generally three orders of magnitude greater than the length of the cantilever (millimeters compared to micrometers), the optical lever greatly magnifies motions of the tip giving rise to an extremely high sensitivity (Braga & Ricci, 2003).

2.2.2.1.4 Image formation

Images are formed by recording the effects of the interaction forces between tip and surface as the cantilever is scanned over the sample. The scanner and the electronic feedback circuit, together with sample, cantilever, and optical lever, form a feedback loop set up for the purpose. The presence of a feedback loop is a key difference between AFM and older stylus-based instruments so that AFM not only measures the force on the sample but also controls it, allowing acquisition of images at very low tip-to-sample forces. The scanner is an extremely accurate positioning stage used to move the tip over the sample (or the sample under the tip) to form an image, and generally in modern instruments is made from a piezoelectric tube. The AFM electronics drives the scanner across the first line of the scan and back. It then steps in the perpendicular direction to the second scan line, moves across it and back, then to the third line, and so forth. As the probe is scanned over the surface, a topographic image is obtained storing the vertical control signals sent by the feedback circuit to the scanner moving it up and down to follow the surface morphology while keeping the interaction forces constant. The image data are sampled digitally at equally spaced intervals, generally from 64 up to 2048 points per line. The number of lines is usually chosen to be equal to the number of data points per line, obtaining at the end a square grid of data points each corresponding to the relative x, y, and z coordinates in space of the sample surface (Braga & Ricci, 2003).

Sahoo, Argawal, and Salapaka (2007) presented a new mode of interrogation at the nanoscale that introduces systems concepts to the area of AFM-based imaging. This method is based on the idea of transient force AFM (TF-AFM) that uses the transient part of cantilever response for imaging (when steady state has not been reached). With the results reported in this journal, it was evident that TF-AFM

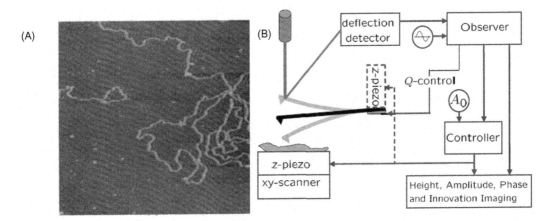

FIGURE 2.7

(A) A typical TF-AFM setup and (B) AFM image taken at scan speed of 4 µm/s.

From Sahoo, D. R., Argawal, P., & Salapaka, M. M. (2007). Transient force atomic force microscopy: a new nano-interrogation method. American control conference. New York: Iowa State University.

can be used to image samples like DNA (which have small feature size) 40 times faster than the conventional speeds.

The DNA samples were prepared as follows: 500 µg/mL Lambda DNA solution was diluted to 50 µg/mL concentration using Tris/HCl/EDTA (ethylene-diamine-tetraacetic acid) buffer (stock solution). Working solution was prepared by further diluting to 1 µg/mL concentration using $NiCl_2$ buffer. For imaging in air, 20 µL of working solution was incubated on a freshly cleaved mica surface for 10 min and then rinsed with pure water and dried using nitrogen gas.

The experimental setup was as follows: TF-AFM required a commercial AFM and an electronic add-on that implemented the observer (Fig. 2.7A). In the current setup, the observer was implemented on an Field Programmable Gate Arrays (FPGA). The innovation signal from FPGA was fed to a true power detector circuit and the output was connected to the auxiliary input of AFM setup. TF-AFM and amplitude modulation mode (AM)-AFM images were then captured in parallel using existing AFM software. Fig. 2.7B shows the DNA image captured by a conventional AFM at a scan speed of 4 µm/s.

As the scan speed was increased to 97.68 µm/s in the conventional AFM, Fig. 2.8A was obtained. Then, by using the TF-AFM with a scan speed of 97.68 µm/s, Fig. 2.8B was captured.

Finally, they took TF-AFM images of DNA at different quality factors of cantilever at scan speed of 97.655 µm/s (Fig. 2.9) and concluded that TF-AFM scan speed is independent on active Q-control.

2.2.3 MAGNETIC FORCE MICROSCOPY

The technique of magnetic force microscopy (MFM) has been discussed extensively in literature, so we will restrict ourselves to a short description. The principle of MFM is very much like that of AFM. Some insist that MFM is just an AFM with a magnetic tip, because in an MFM, much smaller forces are measured. Although this is theoretically true, every MFM is capable of AFM as well (the other way

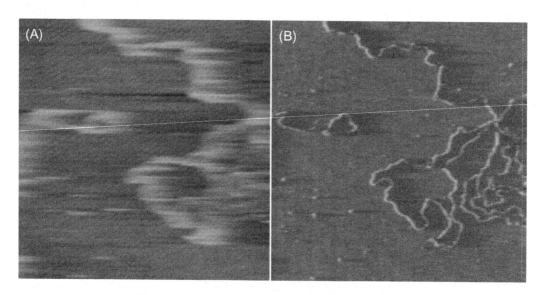

FIGURE 2.8

(A) AFM image taken at 97.68 μm/s and (B) TF-AFM image taken at 97.68 μm/s.

From Sahoo, D. R., Argawal, P., & Salapaka, M. M. (2007). Transient force atomic force microscopy: a new nano-interrogation method. American control conference. New York: Iowa State University.

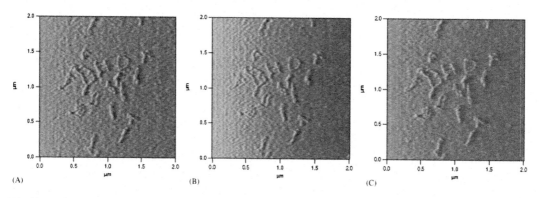

FIGURE 2.9

TF-AFM DNA images at different Q factors. (A) Q = 155; (B) Q = 104; (C) Q = 198

From Sahoo, D. R., Argawal, P., & Salapaka, M. M. (2007). Transient force atomic force microscopy: a new nano-interrogation method. American control conference. New York: Iowa State University.

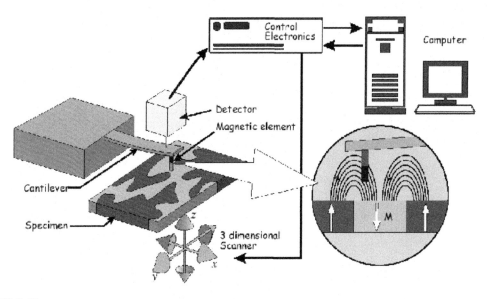

FIGURE 2.10

Schematic diagram of MFM.

From Abelmann, L., Bos, A., & Lodder, C. (2005). Magnetic microscopy of nanostructures. Berlin, Heidelberg: Springer.

around is not generally true). Fig. 2.10 shows a schematic diagram of an MFM. Here, the magnetic stray field above a very flat specimen or sample is detected by mounting a small magnetic element (the tip) on a cantilever spring very close to the surface of the sample.

Typical dimensions are as follows: a cantilever length of 200 μm, a tip length of 4 μm, a diameter of 50 nm, and a distance from the surface of 30 nm. The force on the magnetic tip is detected by measuring the displacement of the end of the cantilever, usually by optical means. The forces measured in typical MFM applications are on the order of 30 pN, with typical cantilever deflections on the order of nanometers. An image of the magnetic stray field is obtained by slowly scanning the cantilever over the sample surface, in a raster-like fashion. Typical scan areas are from 1 up to 200 μm, with imaging times on the order of 5–30 min (Abelmann, Bos, & Lodder, 2005).

Finally, with respect to the fact that MFM can be operated in constant frequency shift mode, two-pass (tapping-lift) mode, or constant height mode, the crucial issue is to minimize surface topography features on the image of the magnetic forces distribution. In order to fix this problem, the majority of MFM measurements are performed in terms of two-pass mode (Hendrych, Kubinek, & Zhukov, 2007).

Hendrych et al. (2007) presented a short historical summary over the MFM technique, which dates back to 1987. Then, they explained that the basic MFM principle comes from the AFM technique, and mentioned that in simple words, the MFM is AFM with the springy cantilever equipped with sharp magnetic probe on its end. Fig. 2.11 shows an MFM setup. As soon as the magnetic probe is brought equally close to the sample surface (units to a hundred nanometers), inconsiderable changes of the cantilever stage caused by mutual magnetic-tip-sample interactions occur (see circle depiction in Fig. 2.11). They can be optically detected (for 2D image the probe is scanned across sample surface).

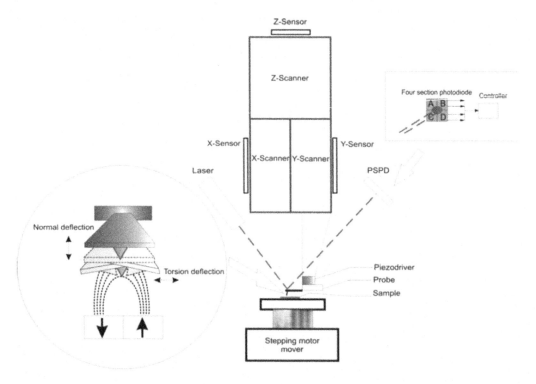

FIGURE 2.11

MFM block controller scheme.

From Hendrych, A., Kubinek, R., & Zhukov, A. V. (2007). The magnetic force microscopy and its capability for nanomagnetic studies—the short compendium. In Modern research and educational topics in microscopy.

Then, they talked about the two MFM operational modes which are static (DC) mode and dynamic (AC) mode and their respective equations. They also briefly introduced the equations for quantitative calculations of the magnetic potential energy and magnetic force acting on the tip.

Finally, they briefly summarized the most noted applications successfully performed by MFM in nanotechnology. These included: (1) new Fe_3B nanowires (Fig. 2.12) and analysis to verify their utility as feasible magnetic composites, MFM tips, or magnetic recording matrices; (2) studies of magnetotactic bacterium in which their cells mineralize magnetic nanoparticles covered by lipid–protein membrane. MFM has evolved from a purely scientific tool to a widely used micromagnetic imaging technique and that this will lead to comprehensive image displaying of various magnetic samples.

2.3 ELECTRON MICROSCOPY

Electron microscopy was first found in Germany in 1931. Electron microscopy is an important and very versatile technique for analysis of nanostructure characterization since it can provide direct images from

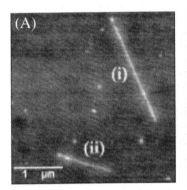 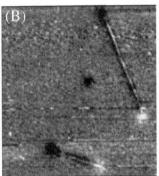 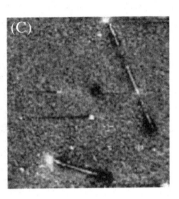

FIGURE 2.12

Fe$_3$B nanowire by (A) AFM, (B) MFM, and (C) MFM after reversion of tip magnetization.

From Hendrych, A., Kubinek, R., & Zhukov, A. V. (2007). The magnetic force microscopy and its capability for nanomagnetic studies—the short compendium. In Modern research and educational topics in microscopy.

which structural details (morphology) can be obtained. Electron microscopy is being used to observe a wide range of biological and inorganic specimens. Electron microscopy utilizes parallel beams of electrons that are accelerated by high voltages and focused through a series of electrostatic or magnetic lenses to illuminate the specimen and produce a magnified image (Martin-Palma & Lakhtakia, 2010).

Electron microscopy includes transmission electron microscopy (TEM) and scanning electron microscopy (SEM). High-resolution (HR) version in electron microscopy which refers to HRTEM and HRSEM are also existing methods, but these two will not be discussed in this chapter.

2.3.1 SCANNING ELECTRON MICROSCOPY

In an SEM, the electron beam is focused over the sample in a manner similar to that used in old-fashioned television sets with cathode-ray tubes. The number of backscattered electrons and/or the secondary electrons generated by the beam that emerge from the sample depends on the local composition and topography of the sample. These electrons are collected by an electron detector, and an image is formed by plotting the detector signal as a function of the beam location. This technique has lower resolution than TEM, typically over 1 nm (Martin-Palma & Lakhtakia, 2010).

A diagram of SEM is shown in Fig. 2.13. The electron gun, which is on the top of the column, produces the electrons and accelerates them to an energy level of 0.1–30 keV. The diameter of electron beam produced by hairpin tungsten gun is too large to form a HR image. So, electromagnetic lenses and apertures are used to focus and define the electron beam and to form a small focused electron spot on the specimen. This process demagnifies the size of the electron source (~50 μm for a tungsten filament) down to the final required spot size (1–100 nm). A high-vacuum environment, which allows electron travel without scattering by the air, is needed. The specimen stage, electron beam scanning coils, signal detection, and processing system provide real-time observation and image recording of the specimen surface (Zhou & Wang, 2006).

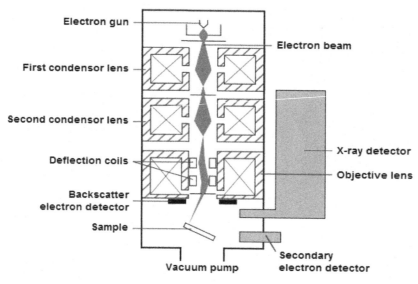

FIGURE 2.13

Schematic diagram of SEM.

*From Steff. (2010, March). File:Schema MEB (en).svg. Retrieved 17.04.11, from Wikipedia: http://en.wikipedia.org/wiki/
File:Schema_MEB_%28en%29.svg.*

Srinivasan et al. (2007) characterized a series of nanoscale chemical patterning methods based on soft and hybrid nanolithographies using SEM with corroborating evidence from STM and lateral force microscopy. SEM was demonstrated and discussed as an analytical tool to image chemical patterns of molecules highly diluted within host self-assembled monolayer (SAM) and to distinguish regions of differential mass coverage in patterned SAMs.

The study has shown that the relative contrast of SAM patterns in scanning electron micrographs depends on the operating primary electron beam voltage, monolayer composition, and monolayer order, suggesting that secondary electron emission and scattering can be used to elucidate chemical patterns. Fig. 2.14 shows the specific images with operating conditions as mentioned in the figure caption.

SEM is used as an analytical tool for the qualitative evaluation of enhanced chemical patterning methods. In particular, the SEM was capable of distinguishing regions of differential mass coverage in patterned SAMs, as well as dilute chemical patterns (<5%) of isolated molecules. The current work also delineates the dependence of the contrast in SEM images of SAMs on operating conditions, which points to the as yet incomplete understanding of the secondary electron emission and scattering mechanisms of molecular overlayers on metal substrates. This indicates the convolution of many parameters, both intrinsic and extrinsic to SAMs, in determining SEM contrast and limits its current application in obtaining quantitative chemical information. However, the SEM is ideally suited to obtain qualitative analytical information such as pattern metrology, with spatial resolution, in chemical patterns of SAMs.

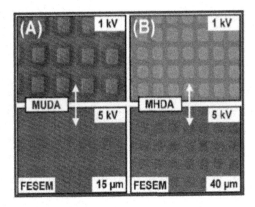

FIGURE 2.14

(A) Field emission scanning electron microcopy (FESEM) images acquired at 1 and 5 kV of microcontact printed regions of 11-mercaptoundecanoic acid (MUDA) that mirror the relief pattern on 10 μm × 10 μm posts on a stamp inked with a 25 mM MUDA (10 min stamp–substrate contact time). (B) FESEM images acquired at 1 and 5 kV of microcontact printed regions of 16-mercaptohexadecanoic acid (MHDA) that mirror the relief pattern on 10 μm × posts on a stamp inked with a 25 mM MHDA (15 s stamp–substrate contact time). The image contrast interchanged reversibly upon switching back and forth between 1 and 5 kV, demonstrating the contrast dependence of SAMs on the operating voltage of the SEM.

From Srinivasan, C., et al. (2007). Scanning electron microscopy of nanoscale chemical patterns. American Chemical Society, 191–201.

2.3.2 TRANSMISSION ELECTRON MICROSCOPY

In TEM, the electron beam travels through the sample and is condensed on a detector plate. A schematic diagram in Fig. 2.15 shows the instrumentation of TEM. Images are formed because different atoms interact with and absorb electrons to a different extent. Since electrons interact much more strongly with matter than do X-rays or neutrons with comparable energies or wavelengths, the best results are obtained for sample thicknesses that are comparable to the mean free path of the electrons (the average distance travelled by the electrons between scattering events). The recommended thickness varies from a few dozen nanometers for samples containing light elements to tens or hundreds of nanometers for samples made of heavy elements. The resolving power of TEM is theoretically subatomic, although resolutions around 0.1 nm have been achieved in practice. Additionally, TEM allows researchers to generate diffraction patterns for determining the crystallographic structures of samples (Martin-Palma & Lakhtakia, 2010).

Yao et al. (2011) described a methodology based on hollow-cone dark-field (HCDF) TEM to study dislocation structures in both nano- and microcrystalline grains. The conventional approach based on a two-beam condition, which was commonly used to obtain weak-beam dark-field TEM images for dislocation structures, was very challenging to employ in study of nanocrystalline materials (especially when grains are less than 100 nm in diameter). Fig. 2.16A and B shows the ray diagram of HCDF-TEM.

The study conducted was on a trimodal Al metal–matrix composite (MMC) consisting of B4C particles, a nanocrystalline Al (NC-Al) phase, and a coarse-grained Al (CG-Al) phase that has been

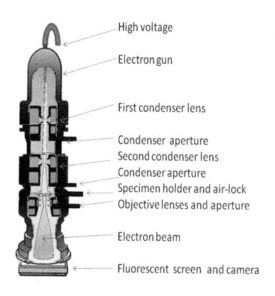

Transmission Electron Microscope

FIGURE 2.15

Schematic diagram of TEM.

From Electron microscope, Wikipedia. https://en.wikipedia.org/wiki/Electron_microscope.

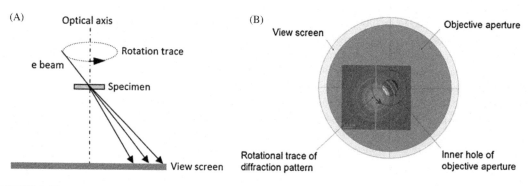

FIGURE 2.16

(A) The ray diagram of HCDF-TEM technique and (B) the diffraction plane configuration for HCDF-TEM imaging.

From Yao, B., Heinrich, H., Smith, C., Bergh, M. V., Cho, K., & Sohn, Y. H. (2011). Hollow-cone dark-field transmission electron microscopy for dislocation density characterization of trimodal Al composites. Micron, 29–35.

reported to exhibit an extremely high strength and ductility. The technique, applied based on HCDF-TEM to examine the dislocation structure, was presented and subsequently employed to examine the dislocation density in trimodal Al MMCs that contain both nanocrystalline (NC-Al) and coarse-grain (CG-Al) AA5083 Al alloy.

The NC-Al phase in the consolidated and extruded sample contained a high density of dislocations ranging from $2 \times 10^{15}\,\mathrm{m}^{-2}$ to $1 \times 10^{16}\,\mathrm{m}^{-2}$. The dislocation density in the CG-Al phase even after hot stretch ratio (HSR) extrusion was measured to be relatively small, $4 \times 10^{14}\,\mathrm{m}^{-2}$. The authors have successfully conducted the examination where they suggest that the CG-Al phase can significantly enhance the ductility of the trimodal Al MMCs.

2.4 DIFFRACTION TECHNIQUES

2.4.1 X-RAY DIFFRACTION

The technique was discovered in 1895 and was then accepted as a characterization method around 1922. X-ray diffraction (XRD) has only been commonly used on nanoparticle characterization during the last 10 years. The main use of XRD is to determine the arrangement of atoms within a crystal of the specimen.

In XRD, a collimated beam of X-rays is directed at the sample, and the angles at which the beam is diffracted are measured. When the beam interacts with an arbitrarily chosen material, its atoms may scatter the rays into all possible directions. In a crystalline solid, however, the atoms are arranged in a periodic array. This periodicity imposes strong constraints on the resulting XRD pattern, so much so that it is possible to determine the crystallographic structure from an analysis of the diffraction pattern. As such, XRD has traditionally been of enormous importance in determining the structures of bulk crystals. In addition to the determination of the crystalline structure of nanomaterials, XRD can provide information on nanocrystal size, microstresses, microstrains, and orientation distribution (Martin-Palma & Lakhtakia, 2010).

Thomas, Loubens, Gergaud, and Labat (2006) focus on the usage of XRD for the need to characterize displacement fields in nanostructures together with the advent of third-generation synchrotron radiation sources which have generated new and powerful methods (anomalous diffraction, coherent diffraction, microdiffraction, etc.). Some of the recent and promising results were reviewed upon the field of strain measurements in small dimensions via XRD. XRD has been used for many years to analyze lattice strains in materials (see Fig. 2.17 for examples). This technique has many distinct advantages compared to other methods.

The combined use of synchrotron radiation with Kikpatrick–Baez optics has enabled the fabrication of highly focused white beams ($<1\,\mu\mathrm{m}$), which are revolutionizing the study of local strain fields. The synchrotron radiation was emitted in a broad energy range allowing the investigation of energy-dependent diffraction. Such anomalous diffraction brings an extra chemical sensitivity to the scattered intensity, which proves extremely useful for those problems where chemistry and elasticity were interrelated. In conclusion, those developments allowed for the implementation of XRD distinct advantages such as full strain tensor determination, analysis of inhomogeneous strain fields, nondestructive nature to nanostructured materials, and provide additional features such as spatial resolution or chemical sensitivity. With the growing need for nondestructive evaluation of strain fields in buried nanostructured materials, more new developments are expected to come in the future.

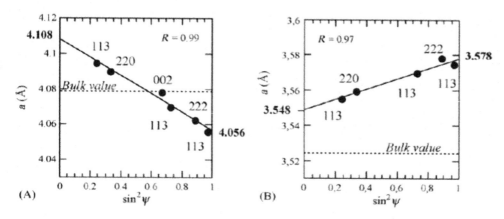

FIGURE 2.17

Typical $\sin^2 \psi$ plot for (A) Au and (B) Ni ultrathin (few nanometer) layers within Au/Ni multilayer.

From Thomas, O., Loubens, A., Gergaud, P., & Labat, S. (2006). X-ray scattering: a powerful probe of lattice strain in materials with small dimensions. Science Direct, 182–187.

2.5 SPECTROSCOPY TECHNIQUES

Spectroscopy techniques are methods that use radiated energy to analyze properties or characteristics of materials. It measures intensity as a function of wavelengths and produces spectrums for comparison purposes.

2.5.1 FOURIER TRANSFORM INFRARED SPECTROSCOPY

Fourier transform infrared (FTIR) spectroscopy is a type of absorbance analyzing technique. It measures the amount of light at different wavelengths that is absorbed by a sample and express these measurements into a spectrum. Since different materials will absorb and transmit different range and levels of light, each will have its unique spectrum. So, by looking at the spectrums, this technique can be used to analyze characteristics such as film thickness, optical characteristics of nanoparticles, optical properties of coating, particle size, and composition.

The technique was first proposed in the early years of the 20th century when Albert Abraham Michelson invented a configuration of beam splitter and mirrors that allows a beam of light to change wavelengths due to interferences (Nicolet, 2001). This component is known as the interferometer in any Fourier transform spectrometer and led to his design being named the Michelson interferometer. It is also the simplest configuration of any interferometer.

The first interferometers designed for this technique became available in the 1960s in England and were marketed by Research and Industrial Instruments Corporation (Griffiths, 1983). It was a slow-scanning instrument that used a paper-tape punch and a remote computer to digitize its results. A faster scanning spectrometer was produced by Block Engineering in the United States afterward. It could detect signals more easily (Griffiths, 1983). The end of the 1960s saw more developments in Fourier spectrometer technology including the Fourier transform algorithm (fast Fourier transform) by Cooley and Tukey in 1965, which enabled higher resolution spectrums to be produced at a faster pace. The algorithm was first implemented in an interferometer in 1969 by Digilab (Griffiths, 1983). Digilab later collaborated with Nicolet to produce a spectrometer that became an industry standard and contributed to the popularity of FTIR spectroscopy today (Griffiths, 1983). Many more improvements have been made since then in various other countries, most of them based on the Michelson interferometer.

This FTIR spectroscopy technique is similar to other absorption spectroscopy techniques, where a light of certain wavelength is passed through a sample. The amount of light transmitted or absorbed is then recorded, and another wavelength is selected to be recorded the same way. The main difference between this method and other absorption spectroscopy methods is its ability to process multiple wavelengths simultaneously with the use of the Fourier transform algorithm. Fourier transform allows the machine to convert raw data of different wavelengths into a spectrum much faster and more efficiently.

There are several components to the FTIR spectrometer: the light source, the interferometer, the sample compartment, and the detector. The light source can contain any wavelength of light from ultraviolet to far-infrared. The FTIR spectrometer gives the most accurate and precise result in the infrared spectrum compared to other methods. Fig. 2.18 shows the general layout of each component. The light source will pass through the interferometer which will adjust the wave to the desired wavelength combination. Then, this light will go through the sample and hit the detector.

The interferometer is the key part of the instrument, using mirrors and a beam splitter in various configurations in order to convert the light source into a beam with different combination of frequencies. The beam splitter is usually a thin film made of plastic that is opaque to wavelengths longer than $2.5\,\mu m$ (Griffiths, 1983).

The most well-known configuration as mentioned earlier is the Michelson interferometer.

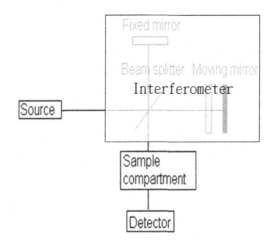

FIGURE 2.18

The general layout of an FTIR spectrometer.

From How an FTIR spectrometer operates, ChemWiki, UCDavis.

Fig. 2.19 shows the Michelson interferometer configuration. It has a single beam splitter in the middle that splits the light source into two equal parts. One beam travels to a fixed mirror and bounces back toward the center and the other beam bounces off a moving mirror. When the two beams join back again in the center, the difference in distance travelled will result in interference between the two waves. The overall effect will produce a light beam of different frequency depending on the distance of the moving mirror.

A more recent interferometer, the refractively scanned interferometer, was discussed in the article, "Fourier Transform Infrared Spectroscopy," by Peter R. Griffiths. The schematic diagram of this configuration is shown in Fig. 2.20.

For this interferometer, a moving wedge beam splitter and two fixed corner mirrors are used. By adjusting the scanning wedge in the center, the ratio between the light pass through to the first mirror and the light reflected to the second mirror will be changed. When the two beams joined back together in the middle, the difference in ratio will create interference resulting in a light of different wavelengths. This configuration is said to exhibit more stability than the Michelson interferometer (Griffiths, 1983).

After exiting the interferometer, the beam is shone through a sample contained in the sample compartment. Part of the radiation will be absorbed while the rest will reach the detector. The detector will then collect the absorbance data, also known as interferograms, and converts them using FT into an absorbance spectrum for ease of comparison with other results.

There are many advantages to using this technique to determine the characteristics of nanoparticles. Compared to other methods, the FTIR spectrometer has a multiplex or Fellgett advantage, which means it is able to analyze information faster because it can process radiation of different frequencies simultaneously. Another advantage is the high quality and precision of its data, made possible because

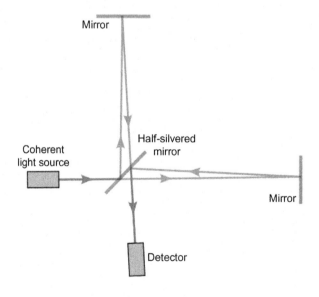

FIGURE 2.19

Schematic diagram of a Michelson interferometer.

From https://commons.wikimedia.org/wiki/File:Michelson_interferometer_with_labels.svg.

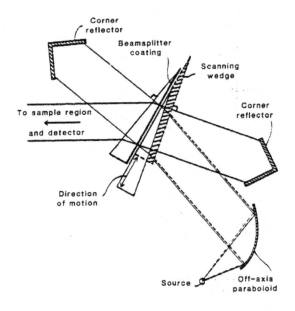

FIGURE 2.20

Schematic diagram of a refractively scanned interferometer.

From Peter R. Griffiths, James A. De Haseth, James D. Winefordner (Series Editor), Fourier Transform Infrared Spectrometry, 2nd Edition. Copyright © 2007, John Wiley and Sons

the FT calculations are done digitally (Griffiths, 1983). A third advantage is the use of interferences in the light, giving this technique a high signal-to-noise ratio which allows it to boost the amplitude of the targeted wavelengths. In addition, remote sensing, which enables the process to not have physical contact with the sample, ensures that the property of the sample remains unchanged.

The FTIR technique has gone through many developments in its instrumentation since the 1960s, mainly to improve the speed of the scanning process. It has provided a new way to test samples using infrared spectroscopy, making analyzing materials "virtually limitless" (Nicolet, 2001). However, there are some limitations to this technique, one of which is that FTIR spectroscopy focuses more on qualitative results as opposed to quantitative (Van de Voort, 2009). As an example, if the FTIR spectroscopy method was used to track relative changes in a substance, the results may be thrown off if the formulation of the substance changes or is covered by another substance.

2.6 SURFACE ANALYSIS AND DEPTH PROFILING

2.6.1 AUGER ELECTRON SPECTROSCOPY

Auger electron spectroscopy (AES) is a characterization technique that is commonly used for the study of material surface. It was first used in 1953 by J.J. Lander and it incorporates a phenomenon called the

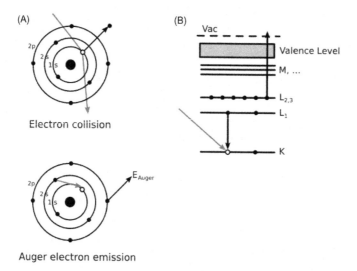

FIGURE 2.21

Two views of the Auger process. (A) It illustrates the Auger deexcitation. An incident electron/photon creates a core hole in the 1s level. An electron from the 2s level fills in the 1s hole and the transition energy is imparted to a 2p electron which is emitted. The final atomic state thus has two holes, one in the 2s orbital and the other in the 2p orbital. (B) It illustrates the same process using spectroscopic notation.

From Auger Effect, Wikipedia. Retrieved from https://en.wikipedia.org/wiki/Auger_effect.

Auger Process to give surface composition information with very high resolution (Grant, 2003). It can also be used to analyze characteristics such as surface roughness, impurities, and binding energies of a sample.

The Auger process was discovered by Pierre Auger in 1925 (Chourasia, 1997). When an atom is exposed to short pulses of X-rays, one of two things can happen: the emission of electromagnetic radiation or the ejection of an electron. The latter is called the Auger Process and the electron is the Auger Electron. As demonstrated in Fig. 2.21, when an atom is exposed to an X-ray pulse or electron beam, an initial electron is ejected to release the extra energy. This causes an empty electron hole in the inner orbital shell. To keep its stability, an outer electron will move in to fill the space and a second electron will be ejected. This second ejected electron is called the Auger electron.

Since different atoms will produce Auger electrons with different energies, AES uses this information to sort the Auger electrons and associate them with their parent atoms to determine the sample's composition.

The instrumentation of AES is usually composed of an electron beam, a sample compartment, an analyzer, and a detector. Instead of X-rays, modern day AES uses electron beams to excite the surface molecules. The beam is usually around 10 nm in diameter, but the smaller the diameter, the higher the resolution produced, because it can analyze smaller area with the same details (Felton, 2003).

As discussed in an article, "On the Surface with Auger Electron Spectroscopy," by Michael J. Felton, there are two main types of analyzers. The analyzer is used to separate the electrons by energy level. The first one is a cylindrical mirror analyzer (CMA). It has a simple configuration, where the sample and electron beam all align in a straight line as shown in Fig. 2.22 (Felton, 2003). When the

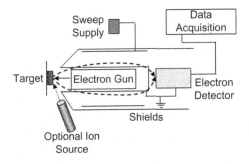

FIGURE 2.22

AES experimental setup using a CMA.

electron gun is focused on the sample, the Auger electrons will be ejected and bounced against the outer mirror, eventually funneled to the analyzer and detector.

The second type of analyzer is the hemispherical analyzer (HSA). It has a more complex alignment, but it can give energy resolution measurement information. Fig. 2.23 is an example of the HSA alignment (Felton, 2003). The electron beam is shot at an angle which allows the Auger electrons to travel through the instrument. Lenses and mirrors are used to focus the electrons so they can reach the analyzer and detector without losing their energy.

Finally, there is the detector in the Auger electron spectrometer. This component is used to count the number of each type of electron after passing through the analyzer. The ongoing trend now is to use multiple detectors (up to 16) within the machine. Less time is required to process and record the data (Felton, 2003). With this data, the types and amount of each atom can be calculated and can give the composition of the area analyzed.

Some advantages of using this technique are that it has high resolution, precise chemical sensitivity, and it is very good in analyzing semiconductors with submicrometer features (Felton, 2003). The precision of the instrument is mainly determined by the diameter size of the electron beam, so AES can be used to analyze surfaces as small as 10 nm.

However, there are some issues researchers are still working on now to improve this technique to make it more efficient. One issue is the backscattering in high-energy electron beam (Felton, 2003). At high energy and small diameter, the beam can sometimes lose focus because some of the ray gets deflected outward. Another problem is that when using high energy, the beam may affect the chemical properties of the sample. Therefore, the current focus now is to create beams that are lower in energy but still high in resolution (Felton, 2003). In addition, improvement on the Auger electron matching software is also useful to the advancement of this technique to make it more efficient.

2.6.2 SECONDARY ION MASS SPECTROSCOPY

The secondary ion mass spectroscopy (SIMS) is a surface analysis technique that analyzes the composition of a sample at the atomic level (around the 100 nm range). It accomplishes this by using a beam

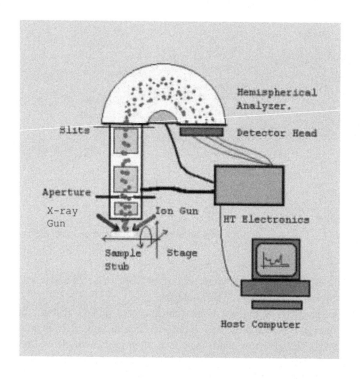

FIGURE 2.23

An example of a spectrometer with a HSA.

From XPS Instrumentation, CasaXPS (2005).

infused with a primary ion to break off molecules from the material's surface. The ejected ions are called secondary ions. These are then used to determine some characteristics of the material such as surface composition, surface impurities, thickness, and 3D profiling.

It started in 1910, when J. J. Thomson first discovered the release of positive ions and neutral atoms when a solid material is bombarded by ions. It was not until the 1960s that SIMS instruments were developed (Benninghoven, 1987). By the year 2002, scientists declared the method no longer a novelty and decided that this technology is at the end of its development. Fortunately, during the following years there were rapid advancements in the field of SIMS due to the incorporation of a charge beam and time-of-flight mass spectrometry (TOFMS) analyzers (Griffiths, 2008).

The SIMS technique works by shooting an ion beam composed of a primary ion, usually argon, gallium, or other alkali metal ions. When the ion hits the surface of the material, its charge and kinetic energy gets transferred over to the surface molecules in contact. This causes those molecules to become ionic and dislodge from the sample. These dislodged molecules are called "secondary ions." In a way, as quoted by Fraser Reich of Kore Technology, "It's like playing molecular pool, the cue ball goes into the material, and it sputters-it lifts off material that's characteristic of the top surface" (Griffiths, 2008). Fig. 2.24 is a pictorial interpretation of this effect.

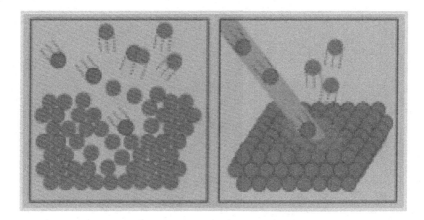

FIGURE 2.24

An ion beam is shot toward the surface of a sample to break it apart.

From https://en.wikipedia.org/wiki/Secondary_ion_mass_spectrometry.

Since each secondary ion will have a specific charge and mass, they can be sorted and counted depending on their mass-to-charge ratios. The result can be used to identify the composition of the material's surface.

SIMS instrumentation involves four components: the ion beam, a vacuum, a mass analyzer, and a detector. As mentioned before, the ion beam is usually composed of a primary alkali metal ion that propels toward the surface of a sample material to slowly chip away its surface for analysis. After the secondary ions are ejected, they are taken through the vacuum toward the analyzer. The analyzer will sort the ions according to their mass-to-charge ratio. When they reached the detector, the amount of each type of ions will be recorded. Fig. 2.25 shows the components of an SIMS as discussed above.

According to Jennifer Griffiths, there are three basic types of mass analyzers: quadrupole, magnetic sector, and time of flight mentioned. The quadrupole analyzer only allows selected mass to pass through and the ions are separated by resonant electric fields. The magnetic sector analyzer uses electrostatic and magnetic properties to separate the ions by their mass-to-charge ratio. The last one, time-of-flight analyzer, separates ions according to their kinetic energy.

Several significant advantages are established using this technique. The SIMS technique has a high sensitivity. Since the area analyzed is determined by the size of the ion beam and ions are very small in diameter, even at low concentration, SIMS is still precise enough to analyze every part of the sample. It also does not require complex sample preparation, which makes it time efficient.

Compared to the early developments of this method, the SIMS today is no longer limited to surface analyzers, as advances have been made in examining samples in three dimensions, known as depth profiling. As the outer layer of the sample slowly chips away, depth profiling allows for a continuous, layer-by-layer analysis of its surfaces. However, it does require a considerable amount of computer data to record every single image of the material in three dimensions, which is still one of its ongoing issues. Other limitations to using SIMS include the physical damage it causes to the sample, allowing only a

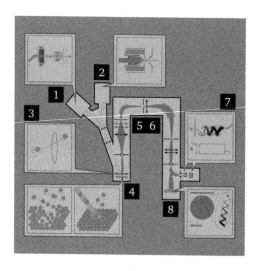

FIGURE 2.25

Schematic of a typical dynamic SIMS instrument. High energy (usually several keV) ions are supplied by an ion gun (1 or 2) and focused on to the target sample (3), which ionizes and sputters some atoms off the surface (4). These secondary ions are then collected by ion lenses (5) and filtered according to atomic mass (6), then projected onto an electron multiplier (7, top), Faraday cup (7, bottom), or CCD screen (8).

From https://en.wikipedia.org/wiki/Secondary_ion_mass_spectrometry.

one-time analysis of the particular sample specimen. In addition, the area analyzed on the sample is dependent on the size of the ion beam.

The future of SIMS is viable as many recent studies have been applied to further this technology. Some of these include the utilization of a cluster ion beam, which uses particles composed of multiple atoms as the primary ion for the spectrometer. With this beam, more complex molecules will be able to break away from the surface. This modification has already propelled SIMS research in the biological field, because scientists are able to study the material surface without reducing it to simple carbons.

2.7 CONCLUSIONS

It is important to analyze the characteristics of nanoparticles in materials in order to understand how they behave, and in turn how they may be beneficial for us to use. There is no one method that ranks over another because each technique has its own advantages and limitations.

Scanning probe techniques such as the AFM and MFM employ excellent instrumentations for producing morphology and micromagnetic imaging, respectively. It may, however, be limited in terms of materials, as the latter can only analyze magnetic ones. Electron microscopy techniques such as the SEM excel at pattern metrology, while the TEM takes extensive sample preparations.

Spectroscopy techniques such as the FTIR allows us to analyze the composition of materials at the nanoscale quickly without harming the sample itself, while the SIMS technique is able to demonstrate 3D profiling of the material. The AES technique meanwhile is good for analyzing surface composition. Overall, these methods all contribute to the production, innovation, and application of nanomaterials.

SMART NANOMATERIALS[*]

3.1 INTRODUCTION

Stimuli responsive or "smart" nanomaterials have been intensively studied in the past 20 years due to growing interest in controlling matter on a smaller scale. There are many areas of application as summarized below (Mansoori, 2005):

Biological:

- Drug discovery
- Antimicrobial (drug-resistant bacteria)
- Biochips (lab-on-chip)
- Targeted therapeutics

Tools and nanodevices:

- Modeling, simulation of nanomaterials
- Nanopositioning, nanomanipulators
- Nanoimaging techniques (STM, AFM, etc.)
- Optical components
- Nano/bio (chemical and biological sensors)
- Nanoelectronics: memory, logic, display, field-emission

Interface phenomena:

- High surface area
- Coating

Synthesis and assembly:

- Self-assembly
- Chemical precipitation
- Physical/chemical aerosol techniques for making clusters/nanoparticles

Materials and molecular machines:

- Use of fullerenes, nanotubes, metals, clays in composite/coating, single-molecule sensing, lab-on-chip, antitumor agent

---

[*]By Yaser Dahman, Adil Kamil, and Daniel Baena.

Nanotechnology and Functional Materials for Engineers. DOI: http://dx.doi.org/10.1016/B978-0-323-51256-5.00003-4

- Materials with grain size less than 100 nm include fabrics, metal powders (used in rocket fuel and skin care products), ceramic fibers, clays, and crystals.

From smart clothing (using cellulose and carbon nanotubes (CNTs)) to drug delivery systems (core–shell particles, hydro and ferrogels, and other colloidal systems), new stimuli responsive nanomaterials are being proposed frequently by scholars and corporations. It is a fairly new field of technology and therefore, there are very few industrial products in the market and endless possibilities for new routes and discoveries. Due to patent restrictions it may take many years for a proposed nanomaterial to go from conception to desired application.

The graph in Fig. 3.1 shows the rise in cumulated patents in all areas of Research and Development of nanotechnology up to 2002. Subsequently, the number of published articles and possible patents increased substantially due to improvements in overall understanding and measurement techniques.

From the graph, "smart" nanomaterials, falling into areas of polymer entrapment, lipid entrapment, and hyperbranched polymers, grew in popularity among researchers around the 1990s. Discussed in this chapter are various "smart" nanomaterials, their response mechanism under different stimuli, some examples of accepted synthesis, and some of their applications.

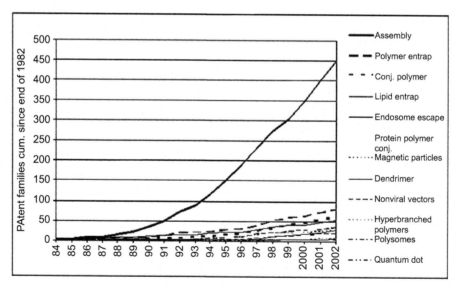

FIGURE 3.1

Nanotechnology patent families. Development of cumulated number of selected patent families in biotechnology between 1984 and 2002.

From Schmidt et al. (2006). Nanotechnology: Assessment and perspectives. Berlin: Springer-Verlag.

3.2 NANOSCALE MATERIALS

3.2.1 HYDROGELS

Hydrogels, also referred to as aqua gels, are a network of hydrophobic polymer chains that are sometimes found as colloidal gels in which water is the distribution medium. Hydrogels can be synthesized via natural or synthetic polymers and experience highly absorbent properties. Furthermore, due to their considerable water content, hydrogels also possess a degree of flexibility analogous to that of natural tissue. Common applications of hydrogels include their use as scaffolds in tissue engineering. Smart environmentally sensitive hydrogels with the ability to sense changes in pH, temperature, or the concentration of metabolite can release their load as a result of such a change. Hydrogels can also be tethered with anticancer drugs and used as a method of sustained-release drug delivery.

3.2.2 POLYMER BRUSHES

Brush polymers are a class of polymers that are grafted or tethered to a solid surface substrate. The polymer brush that is joined to the solid substrate is quite dense in order for the polymer chains to stretch away from the substrate due to volume-excluded effects (crowding among the polymers forces polymers to stretch away from the surface to avoid overlapping). Once the brush polymer is stretched, it exhibits different properties from the original polymer chains in solution. The different properties of stretched brush polymers allow for various applications and uses including its use as novel adhesive materials, protein-resistant biosurfaces, chromatographic devices, and chemical lubricants.

Polymer brushes can occur naturally in nature. Some examples of naturally occurring polymer brushes include extracellular polysaccharides on bacterial surfaces, neurofilaments and microtubules with associated proteins in neuritis, and the proteoglycans of cartilage.

3.2.3 CARBON NANOTUBES

CNTs are hexagonally shaped arrangements of carbon atoms that have been rolled into tubes. Nanotubes are members of the fullerene structural family and are allotropes of carbon. They exhibit extraordinary strength and unique electrical properties and are efficient thermal conductors. They are among the stiffest and strongest fibers known and have remarkable electronic properties. These cylindrical carbon molecules have novel properties, making them potentially useful in many applications in nanotechnology. These properties are shown in Tables 3.1 and 3.2. Nanotubes may be categorized as single-walled nanotubes (SWNTs) or multi-walled nanotubes (MWNTs).

3.2.4 CELLULOSE

Cellulose is a linear organic polysaccharide comprised of many glucose monosaccharides with the chemical formula $(C_6H_{10}O_5)_n$. Cellulose is a hydrophilic substance insoluble in water and most organic solvents. The structure of cellulose is chiral and the compound is biodegradable. It can be broken down chemically into its glucose units by treating it with concentrated acids at high temperature. The characteristic beta acetal linkage in the cellulose structure differs from starch and is the major difference

Table 3.1 Mechanical Properties of Various Engineering Fibers

Fiber Material	Specific Density	E (TPa)	Strength (GPa)	Strain at Break (%)
CNTs	1.3–2	1	10–60	10
HS steel	7.8	0.2	4.1	<10
Carbon fiber PAN	1.7–2	0.2–0.6	1.7–5	0.3–2.4
Carbon fiber pitch	2–2.2	0.4–0.96	2.2–3.3	0.27–0.6
E/S glass	2.5	0/07/0.08	2.4/4.5	4.8
Kevlar 49	1.4	0.13	3.6–4.1	2.8

Table 3.2 Electronic Properties of Various Engineering Substances

Material	Thermal Conductivity (W/mK)	Electrical Conductivity
CNTs	>3000	106–107
Copper	400	6×10^7
Carbon fiber pitch	1000	$2–8.5 \times 10^6$
Carbon fiber PAN	8–105	$6.5–14 \times 10^6$

in digestibility in vertebras. Humans are unable to digest cellulose because the appropriate enzymes to break down the beta acetal linkages are lacking. Cellulose is the main structural component of the primary cell wall of most plants, many forms of algae, and oomycetes. Some species of bacteria secrete cellulose to form biofilms. Cellulose is the most common organic compound on Earth. For industrial use, cellulose is mainly obtained from wood pulp and cotton. It is mainly used to produce paper products such as cotton, linen, and rayon for clothes, nitrocellulose for explosives, and cellulose acetate for films. An immerging use of cellulose is converting cellulose from energy crops into biofuels such as cellulosic ethanol for the study of alternative biofuel sources.

3.3 MECHANISMS OF RESPONSE

The responsiveness of a nanoparticle is a result of a sequence of several events shown in Fig. 3.2: reception of an external signal (either physical or chemical), change of material properties due to signal (pH, thermo, light) or chemical reaction of the material properties (due to interaction with solvents or enzymes), and transduction of the changes into a macro/microscopically significant event (i.e., aggregation or desegregation of nanoparticles, release of cargo) (Motomov, Roiter, Tokarev, & Minko, 2010).

Most "smart" materials contain polymer branches/brushes which allow for response to different stimuli. This section addresses some of these external stimuli.

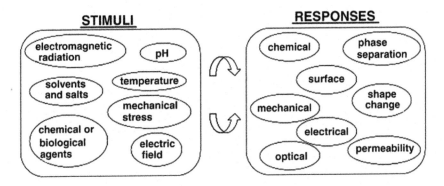

FIGURE 3.2

Potential stimuli and responses of synthetic polymers.

From Schmaljohann, D. (2006). Thermo- and pH-responsive polymers in drug delivery. Elsevier, 1655–1670.

3.3.1 pH-RESPONSIVE NANOMATERIALS

Most pH-responsive polymers are composed of weak polyelectrolytes with carboxylic, phosphoric, or amino functional groups (basic and acidic). By changing the solution's pH, chemical equilibrium is shifted changing the ionization degree (charging neutral particles) of the polymer chains (Ruhe et al., 2004). As a response to this stimulus, the particle expands or collapses its polymer brushes, which is very useful for drug delivery systems in biomedicine (pH throughout the human body differs: 7.35–7.45 for blood and 4.5–5 in lysosomes).

Most pH-responsive materials are core–shell particles produced by bottom–up approaches: chemical synthesis and grafting techniques. The core is usually made up of metals, metal oxides, polymers, etc. Fig. 3.3 is an example of the response of these polymer chains to pH increase.

It is an example of a PVP-based copolymer, in this case poly(2-vinylpyridine). Ionization changes of polyelectrolytes result in coil-to-globule transition. At low pH, the chains retain coil formations which retain mostly the elevation of single chains. As the pH is increased, the polymer chains agglomerate due to intermolecular forces. This agglomeration of polymer brushes is also apparent in hybrid gold–PVP nanoparticles, where these brushes swell and shrink with pH stimulus.

3.3.2 LIGHT-RESPONSIVE NANOMATERIALS

In the case of light-responsive polymers, photoactive groups (azobenzene, spirobenzopyran, or cinnamonyl) are linked to the chains. These light responses are reversible structural changes and the stimulus is usually UV–visible light. Changes in size and shape, or formation of ionic or zwitter-ionic species are outcomes of receiving irradiation (Li & Keller, 2009). Another application of such technology is combining these conjugated polymers with CNTs which provide a good photoactive layer on the surface of the tubes.

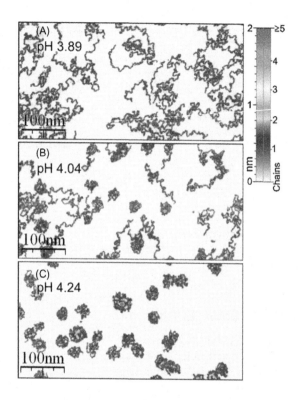

FIGURE 3.3

Changes of the polyelectrolyte ionization result in a pronounced coil-to-globule transition of chains, AFM-visualized conformations of adsorbed poly(2-vinylpyridine). (A) pH 3.89; (B) pH 4.04; (C) pH 4.24.

From Roiter, Y., & Minko, S. (2005). Single molecule experiments at the solid-liquid interface: in situ conformation of absorbed flexible polyelectrolyte chains. Journal of the American Chemical Society, 15688–15689.

An example from Motomov et al. is using amphiphilic-responsive nanoparticles to stabilize foam. By shining light onto the particles, the shell-forming polymer undergoes a chemical reaction by absorption of the light and forms new functional groups. These groups become ionized and change the amphiphilic particles to hydrophobic particles that destabilize the foam.

3.3.3 TEMPERATURE-RESPONSIVE NANOMATERIALS

The response of change in temperature for many nanoparticles varies according to application. Increasing the temperature of a solution can affect the UV–visible absorption spectra of a core–shell nanoparticle, induce changes in wettability of a surface for synthetic and bio-inspired stimulus-responsive systems, and drug release capabilities, similar to "coiled-to-globular" transition in pH (Yusa et al., 2007) (Chen, Ferris, Zhang, Ducker, & Zauscher, 2010).

The coiled to globular effect can be seen in Fig. 3.4, where by increasing the temperature, Yusa et al. (2007) decrease the hydrodynamic radius (R_h) of the poly(N-isopropylacrylamide) (PNIPAM)-based molecule.

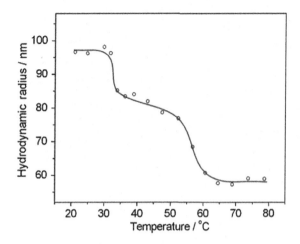

FIGURE 3.4

The hydrodynamic radium (R_h) change of block copolymer PNIPAM-b-PMOEGMA/gold nanocomposites.

From Yusa, S., Yamago, S., Sugahara, M., Morikawa, S., Yamamoto, T., & Morishima, Y. (2007). Macromolecules. 5907.

It is imperative to understand that different polymer chains will act differently under the same stimuli. In some cases, the particle R_h increases with increasing temperature due to balance between segment–segment interactions and segment–solvent intermolecular interactions (Motomov et al., 2010). Examples of polymer–solvent pairs are: PNIPAM, polylactic acid, polylactide (homo- and copolymers), proteins, and polysaccharides in aqueous solutions (Motomov et al., 2010).

3.3.4 MAGNETIC FIELD RESPONSIVE NANOMATERIALS

Magnetic fields (MFs) are used as stimuli for drug release systems. One of the most popular magnetic nanoparticles used in ferrogels is iron oxide (Fe_3O_4). The response here is due to agglomeration and expansion of these magnetic particles within a carrier gel. The mechanism of these responsive materials is summarized in Fig. 3.5.

3.3.5 BIOLOGICAL AND CHEMICAL RESPONSIVE NANOMATERIALS

The response to chemical and biochemical signals is mainly due to interaction between functional groups and chemical reactions within molecules and polymer brushes. This selective interaction relies on colligation of responsive polymers with biological molecules, such as DNA, enzymes, antibodies, and other proteins, and is known as selective molecular recognition phenomena.

The main application of using enzymes as a signaling biochemical is polymers with immobilized enzymes, where a substrate that diffuses from the surrounding aqueous medium to the polymer can be bio-catalytically converted into products which interact with the responsive polymer and cause chemical changes in the polymer and materials with fragments that are substrates for enzymes. Enzyme is used as an "external stimulus" that cleaves the chemical bonds in the material (Motomov et al., 2010).

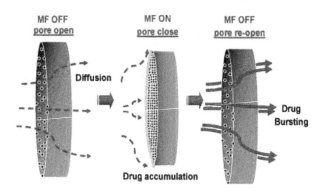

FIGURE 3.5

Mechanism of "on-off" magnetic field.

From Liu, T-Y., Hu, S-H., Liu, T-Y., et al. (2006). Magnetic-sensitive behavior of intelligent ferrogels for controlled release of drug. Langmuir.

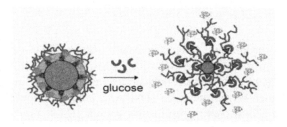

FIGURE 3.6

Schematics of experiment done by Jin et al.

From Jin, X., Zhang, X., Wu, Z., et al. (2009). Amphiphilic random glycopolymer based on phenylboronic acid: synthesis, characterization, and potential as glucose-sensitive matrix. Biomacromolecules.

An example of using enzymes as external stimulus is the experiment done by Jin et al. (Fig. 3.6), where a copolymer made of phenyboronic acid and sugar-based side chains was loaded with insulin. Once free glucose was added, it competed to bind with the boronic acid within the polymer, breaking the cross-links and inducing swelling of the nanoparticle, therefore releasing its payload of insulin.

3.4 SYNTHESIS

There are two major approaches in producing "smart" nanoparticles. The first approach is made of physical processes, based on fabricating particles from presynthesized polymers (cross-linking between polymers and cores). In this section, the coacervation, layer-by-layer, and grafting processes will be discussed. The second approach is a chemical synthesis of nanoparticles by heterogeneous polymerization methods. Below is a detailed discussion of some of the most used synthesis methods of smart nanoparticles. Other examples commonly encountered are as follows:

a. *Adsorption of polymers on nanoparticles*: This is the oldest and simplest method for the fabrication of responsive nanoparticles, where a polymer is adsorbed on the surface of the particle and regulates interactions in the colloidal suspension due to a range of different effects (steric, electrostatic, bridging, and depletion mechanisms).

b. *Self-assembly of micelles and polymersomes from amphiphilic block copolymers*: In this method, block copolymers form various types of self-assembled structures from micelles to continuous bilayers (depending on solvent selectivity). This solvent compatibility results in swelling and packing of particles. Common physical changes within particle are in the aggregate size. Changes to aggregate architecture, structure and responsiveness to pH, ionic strength, thermal, and redox stimuli are among those most commonly considered. Unique examples of stimuli are osmotic shock, shear flow, ionic exchange, etc.

3.4.1 **COACERVATION/PRECIPITATION**

Polymer solutions undergo a liquid–liquid phase separation where the polymer-rich phase is referred to as the coacervate phase. Dispersion of formed colloids is unstable and there is a tendency for coalescence (merging of colloids). However, synthesis allows control of droplet size (either by chemical cross-linking or physical gelation). It is subdivided into simple and complex precipitation. Simple coacervation is often used for entrapping drugs into microcapsules (due to reversibility of process of forming particles), whereas complex coacervation yields particles with two or more stimuli responses. These complex particles are also called "iontropic" hydrogels (Motomov et al., 2010). Another advantage of fabricating using these methods is that particles can be prepared under mild conditions without using organic solvents, surfactants, or steric stabilizers.

Fig. 3.7 shows an example of a pH-responsive nanoparticle prepared by polymerizing an acrylic acid monomer in the presence of gelatin instead of chitosan (Zhang, Wang, Wang, Zhao, & Wu, 2007).

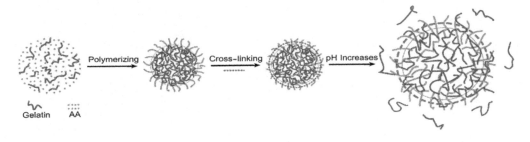

FIGURE 3.7

Schematic of the formation of poly(acrylic acid) (PAA)-gelatin complex coacervate particles and their pH-triggered swelling transition.

From Zhang, Y. W., Wang, Z., Wang, Y., Zhao, H., & Wu, C. (2007). Facile preparation of pH-responsive gelatin-based-core-shell polymeric nanoparticles at high concentrations via template polymerization. Polymer, 5639–5645.

The chemical cross-linking of gelatin with PAA produced a chemically stable complex particle. A surface enriched particle with gelatin chains contributed to colloidal stability, and the swelling/shrinking due to change in pH affected the volume by a factor of 80 (Zhang et al., 2007). Another way of using this process is production from precipitation on templates. This occurs when a solid or polymer particles are present in a solution in which the coacervation process and the polymer-rich phase deposits on the particle surface producing core–shell particles. This phenomenon is also referred to as surface-controlled precipitation and hetero-coagulation of polymers on colloidal particles.

This method allows preparing relatively thick, microscopically homogeneous coatings in a single or a few steps from either charged or noncharged polymers (Radtchenko, Sukhorukov, & Mohwald, 2007). Branches of this process are core–shell and micelle-like particles by complex coacervation, and Micelle-like particles from polyelectrolyte–ionic surfactant complexes.

3.4.2 LAYER-BY-LAYER POLYMERIC SHELL

This is a method of deposition of oppositely charged species (such as organic molecules, proteins, polyelectrolytes, etc.) and was pioneered by Iler (1966) and later studied by Decher Fendler (1997) and Lvov et al. (1995).

It was successfully applied for the fabrication of polymer shells around particulate cores and transformation of the core–shell particles into hollow spheres with layer-by-layer gel walls. An example of this synthesis is the formation of capsules (Fig. 3.8), process design by Donath et al., where a stepwise

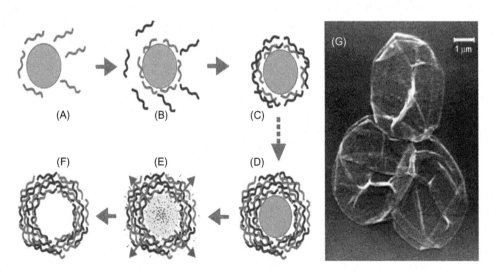

FIGURE 3.8

(A–F) Schematics of PE deposition process and subsequent core decomposition. (G) SEM image of nine-layer [(PSS/PAH)4/PSS] PE shells after solubilization of the MF core. PE, Polyelectrolites; PSS, poly(sodium styrenesulfonate); PAH, poly(allylamine hydrochloride); MF, melamine formaldehyde.

From Donath, E., Sukhorukov, G.B., Caruso, F., Davis, S.A., Helmuth Möhwald, H. (1998). Novel hollow polymer shells by colloid-templated assembly of polyelectrolytes. Angewandte Chemie International Edition.

adsorption was performed of oppositely charged polyelectrolytes onto melanin resin templates (later dissolved in acid).

Steps (A–D) represent the stepwise film formation by repeated exposure of the colloids to polyelectrolytes of alternating charge. The excess polyelectrolyte was removed by cycles of centrifugation and washing before the next layer is deposited. Step (E) was performed after the desired amount of coatings was deposited and the particle was exposed to 0.1 M HCl and the core decomposed immediately. Finally, (F) represents the suspension of free polyelectrolyte hollow shells, poly(sodium styrene sulfonate). By combining weak polyelectrolytes and thermoresponsive polymers in the layer-by-layer shell, a nanoparticle with dual pH and thermoresponsive properties can also be produced (Wang, Cui, Duan, & Yang, 2002).

3.4.3 GRAFTING ONTO THE SURFACE OF PARTICLES

Techniques imply chemical attachment of functional polymers, brushes which are a collection of densely packed polymer chains that provides the particle responsiveness to stimuli. These are normally attached to the core surface by physisorption or covalent chemical attachment (Motomov et al., 2010). Grafting techniques provide nanoparticles sensitive to many stimuli (i.e., changes in solvent quality, pH, ionic strength, temperature).

"Grafting-to" and "grafting-from" are two major approaches. "Grafting-to" is a method that involves a chemical reaction between the presynthesized/end-functionalized polymer brushes and particle surface modified by complementary functional groups. It has advantages such as its simplicity, robustness, and the use of well-characterized polymers (Motomov et al., 2010). Resultant brushes are of moderate grafting density (due to diffusion kinetics of end-functionalized polymer chains) and lead to low film thickness due to excluded volume effects imparted by previously adsorbed polymer chains.

The "grafting-from" method overcomes the limitations of the grafting-to approach, yielding densely packed polymer brushes through surface initiated polymerization. This surface is compatible with a wide range of polymerization chemistries, including anionic, cationic, plasma induced, condensation, photochemical, electrochemical, controlled radical polymerization, and ring-opening metathesis polymerization (Motomov et al., 2010). In the production of "smart" gold nanoparticles (Li, He, & Li, 2009), an initiator immobilized on the surface of gold nanoparticles provides controlled polymer molecular weight, narrow distribution of brushes, and regular shell thickness. Layers of high grafting densities are also produced because the active centers (radicals, ions, and ion pairs) of the growing chains are easily accessible for monomer molecules in the swollen brush in the course of polymerization. A simplified schematic of the difference in synthesis on gold nanoparticles is shown in Fig. 3.9.

3.4.4 HETEROGENEOUS POLYMERIZATION

Many techniques have been developed and have numerous applications for synthesizing monodisperse core–shell particles. Some of the most used synthetic approaches for preparation of "smart" nanoparticles are emulsion, precipitation, and dispersion polymerizations (Pichot, Elaissari, Duracher, Meunier, & Sauzedde, 2001; Pichot, 2004). The main steps of these methods are nucleation and growth.

An advantage is that polymerization allows for synthesis of hybrid particles (organic and inorganic materials—i.e., silica, alumina, iron oxide, and noble metal) allowing for design and production of temperature, pH, ionic strength, and UV-responsive particles. Berndt et al. (Fig. 3.10) synthesized a

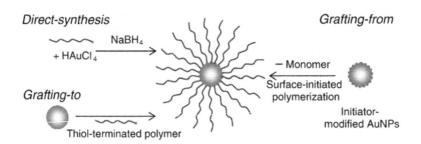

FIGURE 3.9

Simple schematic for producing gold–polymer nanoparticles.

From Li, D., He, Q., & Li, J. (2009). Smart core/shell nanocomposites: intelligent polymers modified gold nanoparticles. Advances in Colloid and Interface Science, 28–38.

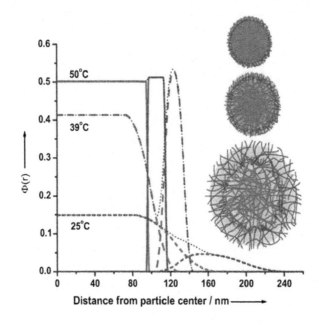

FIGURE 3.10

Radial density profiles for the temperature-sensitive core–shell microgel particles at 25, 39, and 50°C; core (red) (shaded gray in print version), shell (blue) (dark gray in print version), the black dotted lines show the total density.

From Berndt, I., Pedersen, J. S., & Richtering, W. (2006), Temperature-sensitive core–shell microgel particles with dense shell. Angewandte Chemie International Edition, 45, 1737–1741.

microgel composed of a PNIPAM shell and a poly(*N*-isopropylmethacrylamide) in core by applying surfactant-aided radical precipitation polymerization. The difference in the lower critical solution temperature for the polymers was of about 10°C leading to doubly temperature-sensitive behavior.

The synthesized complex particles are responsive in low, high, and intermediate temperatures. At higher temperatures, both shell and core are compressed. In intermediate temperatures, the shell has a higher density than the swollen core, and at low temperatures both shell and core are swollen.

3.5 APPLICATIONS

3.5.1 ENTRAPPING ABILITY AS NANOCARRIERS

Polymer/gold nanocomposites can be used as nanocarriers for drugs, catalysts, and biosensors because the polymer scaffold can be used to immobilize smaller nanoparticles or biomolecules (Li et al., 2009). In the stimuli-responsive process, switching the polymer layer from a hydrophilic state to a hydrophobic state results in intake–outtake water behavior. The process can entrap small nanoparticles or biomolecules into the polymer network. In addition, polymer/gold nanocomposites can be soluble in different solvents due to the solubility of outer polymer layers, which provide a possibility of drug delivery between immiscible water/oil phases (Li et al., 2009).

Such dispersible capability of the polymer/gold nanocomposites in water and organic solvents allows themselves to be used as nanocarriers for drug entrapment and release in different surroundings. The outside polymer network around gold cores can also immobilize the biomolecules such as enzyme, DNA, and so on to fabricate biosensors and bioactuators (Li et al., 2009).

3.5.2 BIOLOGICAL POTENTIAL

The fact that gold nanoparticles typically stimulate low toxic side effects has made the use of polymer-modified gold nanoparticles appealing in medical applications (Li et al., 2009). For example, studies show that PEG shells can greatly diversify the nanoparticles properties and potential applications in areas such as organic solubility, water solubility, and biocompatibility (Li et al., 2009). Research has shown that the use of methoxyl-PEG-thiol or coumarin-PEG-thiol functionalized gold nanoparticles demonstrate anti-biofouling properties and can be used as tomography (CT) contrast agents with higher efficiency (Li et al., 2009).

Furthermore, polyethylenimine-modified gold nanoparticles can be used as gene delivery devices and were found to transfect monkey kidney (COS-7) cells six times better than polyethylenimine alone (Li et al., 2009). The polyethylenimine–gold nanoparticles complex improved cell transfection by increasing the effective molecular weight of the gene delivery system, this resulted in enhanced DNA binding and condensation. Furthermore, the development of intelligent therapeutic devices which have the ability to merge factors such as molecular recognition, biocompatibility, and intelligent response can be synthesized from thermally responsive polymer/metal nanocomposites (Li et al., 2009). For instance, the temperature-responsive, polymer-modified gold nanoparticles can be used for drug delivery applications. The polymer–gold nanoparticle complex can encapsulate the desired drug compounds, the gold nanoparticle core exposed to visible-to-near infrared wavelengths of light subsequently absorbing the light and converting the energy into heat (Li et al., 2009). This increase in heat will result in the swelling of the polymer shells which is an ideal change in drug delivery applications.

3.5.3 **FIELD-EFFECT TRANSISTORS**

The unique electrical and mechanical properties of CNTs can be utilized in order to construct novel transistors and semiconductors (Li et al., 2003). By varying the diameter of the CNT, it is possible to control the band gap of semiconducting nanotubes, because the width of the diameter of the nanotube is inversely proportional to the size of the band gap (Li et al., 2003). Semiconducting nanotubes can be used to build molecular field-effect transistors, while metallic nanotubes can be used to build single-electron transistors. Recent research has developed SWNT transistors that are capable of operating under room temperature.

3.5.4 **FIELD EMISSION DISPLAYS**

The extraordinary electrical properties of CNTs make them ideal candidates for next-generation display devices (Li et al., 2003). The emission of electrons from a wire induced by external electromagnetic fields is the primary concept of a field emission device (FED). The efficiency of generating electrons is dependent on the strength of the electric field and the geometry of the wire (Li et al., 2003). The electromagnetic field amplification increases with increase in electric field and with a decrease in wire radius. CNTs exhibit strong electric fields that are 100 times the electric field of conventional FED material at a diameter of ~1 nm.

The structure of an SWNT-based flat panel display is shown in Fig. 3.11. Each pixel in the flat panel device is made of an anode glass which is coated with a layer of indium-tin-oxide phosphor and a cathode glass that is coated with SWNT. The role of the SWNT in this flat panel device is to generate an electron source. By assembling these SWNT pixels into a matrix a flat panel display is achieved.

FIGURE 3.11

The Samsung 4.5″ full-color nanotube display.

From Paradise, M., Goswami, T. (2007). Carbon nanotubes—Production and industrial applications. Materials and Design 28, 1477–1489—Dr. W. Choi of Samsung Advanced Institute of Technologies.

3.6 **CURRENT RESEARCH AND FUTURE PERSPECTIVES**

The topic of "smart" nanotechnologies branches into a variety of fields, from nanomedicine to electronics. Important to many of these applications are core–shell responsive nanoparticles. It is thus important to understand their synthesis and responsiveness to external stimuli. Li et al. (2009) discussed different methods for preparing polymer-modified gold nanoparticles and their thermo-sensitivity and pH-responsiveness. The applications of such smart nanocomposites vary from: (1) nanocarriers for catalysts and biosensors (due to "live" outside polymer shells), (2) novel 2D/3D nanostructures for use in nanoelectronics, spintronics, and nanosensors (due to ability of self-assembly at surfaces or interfaces), and (3) intelligent pharmaceutical nanosystem with a controlled fashion for biological purposes. A direct synthesis yields nanocomposites with relatively good dispersity, long-term stability, and provides gold cores with a broad size distribution. Covalent "graft-to" strategy prepares gold nanoparticles with sulfur terminated polymers and yields low polymer graft density (due to the steric hindrance from large polymer chains). The "graft-from" strategy on the other hand provides controlled polymer molecular weight, narrow distribution, and shell thickness regularity.

According to Li et al.'s work, thermoresponsive materials, such as PNIPAM-b-PMOEGMA, show significant decrease in apparent R_h (due to contraction of polymer brushes) as the temperature increased; *2-phenylprop-terminated PNIPAM* molecules at high temperature aggregate. Fig. 3.12 shows transmission electron microscopy (TEM) images of gold nanoparticles and polymer brushes in different pH solutions. It was found that for polymers such as PVP at higher pH levels, these brushes would contract and individual particles agglomerate. As pH decreases, these intramolecular forces decrease, spreading out the molecules and expending the particle's polymer brushes. The study concluded that these smart nanocomposites comprising of intelligent polymers and gold nanoparticles can be prepared for different applications (each synthesis yields certain properties). The responsive polymer brushes undergo hydrophilic–hydrophobic phase transitions in response to stimulus such as temperature and pH differences.

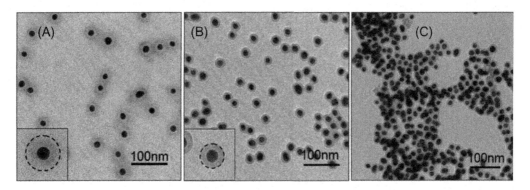

FIGURE 3.12

TEM images at pH 3.1 (A), pH 4.4 (B), and pH 7.5 (C) in a detailed study of the PVP/gold nanoparticles.
From Li, D., He, Q., & Li, J. (2009). Smart core/shell nanocomposites: intelligent polymers modified gold nanoparticles. Advances in Colloid and Interface Science, 28–38.

In the field of nanomedicine, drug delivery systems have been a central topic of study. Biocompatibility of materials and innovative new ways of self-discharging drugs both hinder and allow for a wide range of mechanisms. There are a variety of mechanisms for drug delivery. Liu et al. (2009) studied the effect of particle size, "on-off" MF modes, and switching duration time (SDT) on ferrogels synthesized through freezing–thawing cycles combining poly(vinyl alcohol) hydrogel and magnetic particles (Fe_3O_4, Fig. 3.13A) of diameters ranging from larger particles (LM, 150–500 nm), middle size (MM, 40–60 nm), and smaller magnetic particles (SM, 5–10 nm). An MF was applied to PVA5-LM17, and it was observed that once ON, the swelling ratio of the ferrogel decreased (Fig. 3.13B, -▲-). This resulted in a decrease of drug release (Fig. 3.13B, -○-), since the particles contract and impede the diffusion of drug through the gel, "locking in" any drug within the magnetic particles. The permeability coefficient of PVA5-LM17 when the MF is "off" is $586 \times 10^{-6}\,cm^2/min$. Once switched "on," the coefficient decreases sharply to $40 \times 10^{-6}\,cm^2/min$.

The SDT also affects the release profile of the drug, it was found that for a 10 min cycle, the "close" configuration (which interlocks particles and reduces permeation of drug) was most effective in diffusing the drug once in the "off" mode and had a sharp decrease once returned to the "on" mode. LM particles were proven to be most suited for this experiment, since it presents the most "magnetic sensitive effects." Smaller particles tended to aggregate together and therefore affected the "close" configuration. Ferrogels made of medium size particles had poor drug permeation. The study concluded that "magnetic sensitive effects" are best observed in PVA5-LM17 ferrogels, because of superior saturation magnetization allowing strong MF induction, and smaller coercive force indicating ability of reorientation of particles within the gel under an MF.

Much work is being done on innovative ways to improve or change drug delivery systems. Wu et al. (2010) worked on a single nanoparticle that can integrate optical pH-sensing, cancer cell imaging, and controlled drug release. They discussed the characteristics and applications of an HPC–PAA–MBAAm

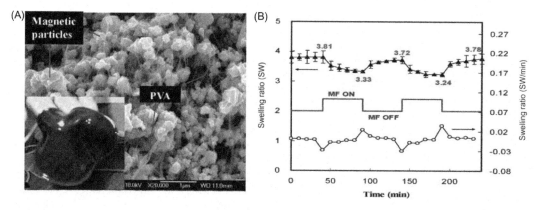

FIGURE 3.13

(A) Cross-sectional SEM image of magnetic particles disperse in PVA hydrogels and OM photos of PVA5-LM17 ferrogels and (B) swelling ratio and swelling rate of PVA5-LM17 ferrogel in the MFs switching "on–off" mode.

From Liu, T-Y., Hu, S-H., Liu, T-Y., et al. (2006). Magnetic-sensitive behavior of intelligent ferrogels for controlled release of drug. Langmuir.

nanogel developed by the group with in-situ immobilization of CdSe quantum dots (QDs). The study examined the gel's pH-sensing, tumor cellular imaging, and pH-regulated drug release. Each element of the gel served a purpose: cadmium selenium QDs were used for optical identification codes for sensing and imaging, HPC chains'-OH groups prevented agglomeration and release of QDs, PAA chains provided pH sensitivity and allowed for delivery systems via swelling/shrinking transition, and carboxyl groups were used for bioconjugation and potential targeting ability. This hybrid nanogel is able to emit different fluorescence wavelengths, which is beneficial for sensor and bioimaging applications. Its ability for carrying drugs (such as TMZ) and high drug loading capacity was studied under different pH stimuli and was found to be very effective. Fig. 3.14 shows the two template nanogels (without the 3.2 nm QDs). Denser cross-links in HNG-2 resulted in smaller swelling degree under different pH solutions affecting drug release properties of the hybrid gel.

Throughout their experiment the stability of the nanogel was studied and even at the most swollen state, QDs were not released from the nanogels due to the strong binding. The study concluded that the developed hybrid nanogel shows excellent ability as a drug carrier due to its porous nature and swelling abilities, providing opportunities for combined diagnosis and therapy ability (switch on and off of nanogel functions). Also, the addition of phosphorescent QDs enhances optical features, which are beneficial for sensor and bioimaging applications.

As discussed in Chapter 4, Nanosensors, sensors are an important part of nanotechnology. The use of smart nanomaterials can be incorporated into the field of strain sensors, as illustrated by Kuilla et al.'s (2010) study on the use of graphene-based piezoresistive smart nanomaterials. Kuilla et al. examined graphene-based nanohybrid material and its piezoresistive characteristic in order to develop smart novel materials for potential application in graphene strain sensors. A graphene/epoxy composite was fabricated

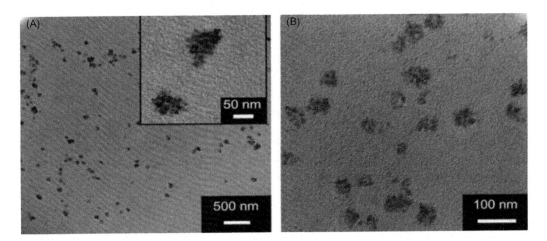

FIGURE 3.14

TEM images of HPC/AA/MBAAm template stable nanogels (A) the HNG-1 (40:10:10wt%) and (B) HNG-2 (40:15:20wt%) R_h <100nm.

From Wu, W., Aiello, M., Zhou, T., Berliner, A., Banerjee, P., Zhou, S. (2010). In-situ immobilization of quantum dots in polysaccharide-based nanogels for integration of optical pH-sensing, tumor cell imaging, and drug delivery. Biomaterials.

via casting mold. Graphene was directly distributed into an aqueous epoxy polymer without the use of a solvent via an ultrasonic homogenizer for 2 h. Jeffamine polytheramines was added in order to induce curing. The subsequent mixture of graphene, epoxy, and jeffamine was placed in a vacuum oven at 50°C and 70 cm Hg for 10 min. The grapheme/epoxy composite was then poured into a silicon mold and cured at room temperature for 24 h. After removing the mold, the cast composite was cut into the desired sizes. In order to fabricate a sensor, two electrodes were connected onto the cast composite with silver conducting epoxy. The sensor was attached to a steel beam. The piezoresistance of the sensor was measured via a multimeter. The beam deflection was measured with a laser displacement sensor to find the sensor deformation. A cantilever beam was used to model and test the sensor. The beam displacements and the subsequent sensor resistance were measured. The sensor results were compared with composite CNTs strain sensors. The strain response of graphene/epoxy sensor was found to be symmetrical and exhibit reversible behavior. The gauge factor obtained was approximately 11.4 and within a range of 1000 microstrain. The graphene composites showed higher gauge factor than strain gauge made of high-quality graphene film. Furthermore, the strain sensitivities of graphene composites were found to be much higher than CNT composites. It was concluded that higher strain sensitivity of graphene composites may have been due to the larger inter-contact areas among the graphene nanofillers due to their 2D structure.

One of the emerging applications of smart nanomaterials is in the development of smart nanotextiles. The synthesis of these smart textiles often requires the use of sophisticated techniques. One such technique was outlined in the research conducted by Kinloch, Li, and Windle (2004). Kinloch et al. examined the spinning of CNT fibers and ribbons directly from the chemical vapor deposition (CVD) synthesis zone of a furnace using a liquid source of carbon and an iron nanocatalyst. In this process, liquid ethanol (carbon source) was used in which 0.23–2.3 wt% ferrocene and 1–4 wt% thiophene were dissolved. The solution was then injected from the top of a furnace into a hydrogen carrier gas. The furnace temperature was maintained between 1050°C and 1200°C. The high temperature vaporized the ethanol/ferrocene/thiophene complex and the process underwent CVD to form an aerogel of nanotubes. The aerogel is captured and wound out of the hot zone continuously as a fiber by a rotating spindle. The alignment, purity, and structure of the fibers obtained from the ethanol-based reactions were characterized by electron microscopy, including image analysis, Raman spectroscopy, and thermogravimetric analysis. The quality of alignment of the nanotubes was measured from transforms of scanning electron microscope (SEM) images. The formation of MWNT or SWNT can be controlled by adjusting the reaction conditions. If the thiophene in the ethanol feedstock is adjusted to between 1.5 wt% and 4.0 wt% and the reactor temperature is between 1100°C and 1180°C MWNTs are formed, while if the thiophene concentration is reduced to 0.5 wt% with a reaction temperature of up to 1200°C SWNTs are formed. It was observed that the purity of the CNTs fibers were directly proportionate to the hydrogen flow rate. High hydrogen flow rates were found to suppress the formation of other forms of carbon impurities. When analyzing the MWNT fibers, the nanotubes diameters were 30 nm, with an aspect ratio of ~1000. They contained 5–10 wt% iron but no other carbon contaminants. It was observed that the degree of alignment can be improved if greater tension is applied to the fiber during processing. The SWNT fibers contained more impurities than the MWNT fibers, with the proportion of SWNTs estimated from transmission electron microscope observations as being more than 50 vol.%. The SWNTs had diameters between 1.6 and 3.5 nm and were organized in bundles with lateral dimensions of 30 nm. The purity of MWNTs in fibers spun at high temperatures was higher (85–95 wt% purity) than for material collected from the furnace without spinning (70–85 wt% purity) (Fig. 3.15).

As previously mentioned, one of the criteria of smart material is their ability to react to changes in their surrounding environment. There has been a growing interest in synthesizing smart nanomembranes

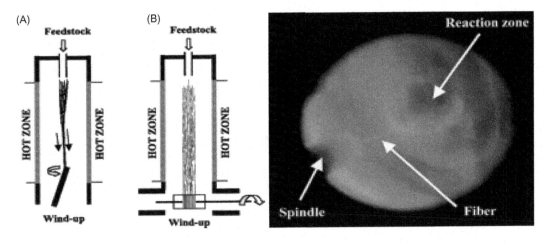

FIGURE 3.15

(Right) Depiction of CNT fiber being drawn on rotating spindle. (Left) A video frame view up the furnace, showing the nanotubes being drawn from the aerogel into the fiber on the spindle.

From Kinloch, I., Li, Y., & Windle, A. (2004). Direct spinning of carbon nanotube fibers from chemical vapor deposition synthesis. Science Magazine, 276–278.

with the ability to change their permeability when subject to external stimuli. Such material can have great potential in medicine in the case of drug delivery systems. The ability to synthesize such smart membranes was researched and the subsequent results were published by Csetneki et al. (2006). Csetneki et al. examine novel composite-gel membranes containing nanochannels that are capable of regulating membrane permeability in response to external temperature change. The channels contain an ordered array of magnetic polystyrene latex particles that undergo change in volume in response to external stimuli. The magnetic polystyrene latex was prepared by mixing ferrofluids of magnetite (Fe_3O_4) particles with an average diameter of 10 nm with a mixture of styrene, sodium lauryl sulfate, stearyl alcohol, and N,N'-azobis (isobutyronitrile). Magnetic polymer latex was prepared by the miniemulsion technique and was allowed to polymerize. The latexes were subsequently subjected to water-vapor distillation and washed to remove the unreacted monomers. N-isopropylacrylamide monomer, methylene bis acrylamide as the cross-linker, and potassium persulfate (KPS) as the initiator were mixed with the magnetic polystyrene latex. The shell polymerized under a nitrogen atmosphere resulting in particles of core–shell MPS–PNIPAM microgel latex with a thermosensitive PNIPAM surface layer. The membrane was fabricated by mixing the MPS–PNIPAM microgel latex with a solution of polyvinyl alcohol (PVA) containing glutaric aldehyde as crosslinking agent. The cross-linking reaction of PVA was induced by adding a few drops of hydrochloric acid. The mixture was transferred into a mold and subsequently placed perpendicularly to a uniform MF. The resulting reaction locked and aligned the chainlike structure in the gel along the direction of the MF. The core–shell MPS–PNIPAM particles formed channels in the PVA matrix and when subject to temperature change began to swell. It was concluded that as the temperature surrounding the membrane changed, the permeability drastically changed. With increasing temperature the permeability increased. Below the critical temperature (37°C) the channels in the PVA membranes were saturated with MPS–PNIPAM latex beads and as a result the solutes permeability was limited. The polymer chains restrict the mobility of the

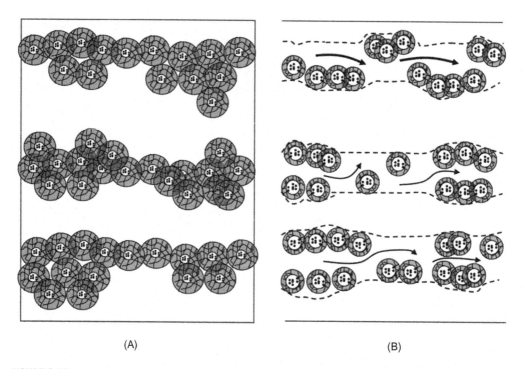

(A)   (B)

FIGURE 3.16

Schematic representation of channels made of MPS–PNIPAM latex built in the PVA gel matrix: (A) "off" state below the critical transition temperature and (B) "on" state above the critical transition temperature.

From Ildiko, C., Genoveva, F., & Miklos, Z. (2006). Smart nanocomposite polymer membranes with on/off switching control. Macromolecules, 1939–1942.

solute molecules, resulting in a reduced diffusion coefficient. When the temperature was increased above the critical transition temperature (37°C), the size of the MPS–PNIPAM latex beads reduce causing the bead diameter to be less than that of the channel, the MPS–PNIPAM beads no longer filled up the entire channels, and polymer free cavities were formed. The open cavities allowed for an increase in solute permeability and an increase in diffusivity (Fig. 3.16).

3.7 CONCLUSIONS

Currently, nanotechnology has influenced improvements in smart materials by implementing relatively simple changes to existing technologies. However, the future holds possibilities for solutions to seemingly impossible problems, such as combating disease, development of novel technologies, and solving the global fuel crisis. The solutions to these problems will inevitably incorporate smart materials such as nanosensors, nanocomputers, and nanomachines into their structures. The advent of smart nanotechnology has the potential to greatly improve every aspect of human interaction.

NANOSENSORS[*]

4.1 INTRODUCTION

The purpose of sensors is driven by our need to monitor and receive information about the environment surrounding us. Sensors transduce information in the form of variations of energy, such as thermal, mechanical, optical, electrical, magnetic, or biochemical into another form of energy. For example, the information about the kinetic energy in a vehicle could be revealed by the increase of temperature due to the heat released by a disk brake. The advancements in the field of nanotechnology over the past few years have been astounding and the field of nanosensors is no exception to this. A nanosensor is a physical, biological, or chemical sensor that is built on the atomic scale in measurements of nanometers (i.e., 10^{-9} m) used to convey information about nanoparticles to the macroscopic world. Generally, a nanosensor is defined as a sensor fulfilling at least one of the three following requirements (Wiesendanger, 1995):

- The sensitivity of the sensor is on the nanoscale (e.g., sensors for displacements on the nanometer scale, force sensors with sensitivity of the order of nanonewtons, power sensors with sensitivity of the order of nanowatts).
- The spatial confinement of the interaction of the sensor with the object is on the nanometer scale.
- The size of the sensor is on the nanometer scale.

Nanosensors have been under research by many institutions, with great advancements in the development of nanosensors for many different applications within medical, national security, aerospace, and integrated circuits fields. In general, engineered nanoscale materials demonstrate many unique and desirable physical properties such as increased reactivity, optical absorption, catalytic efficiency, electrical conductivity, wear resistance, strength, and magnetic properties in comparison to bulk matter of the same composition (Riu, 2005). In comparison to existing sensor technologies, nanosensors are small and lightweight, have a large reactive surface area, greater selectivity, improved sensitivity, and improved response times. One of the major advantages of downsizing to the nanoscale is the massive increase in surface area. For example, a single-walled carbon nanotube has a surface area of $1600 \, m^2/g$. To put this into perspective, just 4 g of nanotubes have approximately the same surface area of a football field (Shelley, 2008). Since many sensors are dependent on surface chemistry, a larger surface area provides a great advantage to nanosensors wherein an increase in signal intensity is possible allowing for trace amounts to be detected.

---

[*] By Yaser Dahman, Amira Radwan, Boris Nesic, and Jason Isbister.

Nanotechnology and Functional Materials for Engineers. DOI: http://dx.doi.org/10.1016/B978-0-323-51256-5.00004-6

4.2 MANUFACTURING METHODS

Nanosensors can be manufactured in a number of methods, but three of them are most common. These methods are top–down lithography, bottom–up assembly, and molecular self-assembly.

4.2.1 TOP–DOWN LITHOGRAPHY

The most common top–down approach to nanosensor fabrication involves lithographic patterning techniques using short-wavelength optical sources. In the top–down approach, highly ordered nanolines can be obtained via different lithography methods such as electron beam lithography or nanoimprint lithography. The term "lithography" refers to a top–down fabrication technique where a bulk material is reduced in size to a nanoscale pattern in the order of about 25 nm (Vazquez et al., 2008). A key advantage of the top–down approach is that the parts are both patterned and built in place, so that no assembly step is needed. Lithography involves a number of related processes, like resist coating or exposure. Generally, the method entails beginning with larger blocks of material and carving them out into the desired form. The pieces that are carved out are then used as components to the specific microelectronic systems, such as the nanosensors. This method is used to create integrated circuits. In the case of nanosensors, a uniform layer of photoresist is first deposited onto the surface of a silicon wafer. The photoresist, with thickness typically from a few nanometers to a micrometer, is deposited by spin coating. Selected areas of the resist are subsequently exposed to a radiation source through a mask. Upon sufficient exposure, the polymer chains in the resist are either broken or cross-linked leaving a positive or negative resist, respectively. The piece of material can then be doped and modified using other materials that can produce nanosensors of various functions.

Since the lithography process transforms a 2D pattern into a 3D structure in the resist and the silicon wafer, the depth profiles in both materials are very important. By selecting a suitable developer, temperature, and developing time, one can obtain different tailored profiles in the resist. The pattern transfer can be realized in two general processes: from the photoresist to the silicon wafer by etching or by postdeposition onto the patterned photoresist by lift-off or electrodeposition (Vazquez et al., 2008). Since the lithography resolution is determined by the radiation wavelength, lithography is often categorized by the radiation source, namely—optical, electron-beam, ion-beam, or X-ray lithography. In optical lithography, UV light is usually used ($\lambda = 193$ nm). In electron-beam lithography, an electron-sensitive resist is exposed to an electron beam through scanning electron microscopy (SEM) or transmission electron microscopy. In the case of X-ray lithography, the key point is the exposure of a resist to X-ray radiation in a parallel replication process. Similar to the electron-beam lithography, the sample is covered by a resist layer with high sensitivity in the X-ray wavelength form. The combination of electron-beam lithography with electroplating techniques enables the fabrication of patterned elements of high aspect ratios, such as nanowires. Fig. 4.1 illustrates the SEM image of an array of Ni nanowires fabricated using combined electron-beam lithography and electrodeposition techniques.

4.2.2 BOTTOM–UP ASSEMBLY

The opposed direction of the top–down assembly approach is the bottom–up approach, where molecules and atoms are assembled into structures up to the nanometer scale. The bottom–up approach is a promising nanotechnology alternative to the top–down methods that were inherited from the

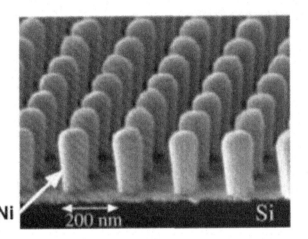

FIGURE 4.1

Ni nanocolumns prepared by combined electron-beam lithography and electrodeposition.

From Vazquez, M., Asenjo, A., Morales, M. d., Pirota, K. R., Confalonieri, G. B., & Velez, M. H. (2008). Nanostructured magnetic sensors. In F. J. Arregui, Sensors based on nanostructured materials (pp. 196–198). Pamplona, Spain: Springer.

macroworld (Andrue et al., 2008). For nanosensors, the bottom–up assembly approach uses atomic sized components as the basis of the sensor. These components are moved into position one by one in order to create the nanosensor. It should be noted that this method poses extreme challenges for mass production, since bottom–up assembly has only been used in laboratories with the use of atomic force microscopy.

Over the past years, top–down and bottom–up approaches have overlapped in order to facilitate and accelerate the discovery of new functional materials. However, the bottom–up approach enables the design and synthesis of novel materials with specifically tailored properties for the sensing processes that cannot be imagined from top–down approaches or any conventional microtechnologies. Due to specifically tailored properties, detection limits can be lowered and sensitivity can be increased. The bottom–up approach is driven mainly by the reduction of Gibbs free energy, so the nanomaterials thus produced are in a state closer to thermodynamic equilibrium (Varadan, Chen, & Xie, 2008). The synthesis of large polymer molecules is a typical example of the bottom–up approach, where individual building blocks, monomers, are assembled into a large molecule or polymerized into bulk material. Crystal growth is yet another example, where atoms, ions, or molecules assemble in an orderly fashion into the desired crystal structure on the growth surface (Varadan et al., 2008).

Nanostructures fabricated through the bottom–up approach usually have fewer defects, a more homogeneous chemical composition, and better short- and long-range ordering. However, one of the main challenges of the bottom–up approach is fabricating structures that are of sufficient size and amount to be used as materials in practical applications. For this reason, some applications are developed through a hybrid approach of bottom–up and top–down. For example, a magnetic sensor is usually developed through a hybrid approach, since the thin film is grown through a bottom–up approach, and etched into the sensing circuit through a top–down approach.

4.2.3 **MOLECULAR SELF-ASSEMBLY**

Molecular self-assembly is a type of bottom–up approach by which molecules adopt a defined arrangement without guidance form an outside source. Using molecular self-assembly, the desired structure is programmed in the shape and functional groups of the molecules. Molecular self-assembly has a number of advantages. First, it carries out many of the most difficult steps in nano-fabrication (those involving atomic-level modifications) using very highly developed techniques of synthetic chemistry. Second, it draws from the enormous wealth of examples in biology for inspiration, since self-assembly is one of the most important strategies used in biology for the development of complex, functional structures. Third, it can incorporate biological structures directly as components in the final systems. Finally, because it requires that the target structures be the most stable thermodynamically, it tends to produce structures that are relatively defect-free and self-healing (Whitesides, 1995).

There are two methods to the concept of molecular self-assembly which is also known as growing nanostructures. The first method is to use a piece of previously created or even naturally formed nano-structure as the base and immersing it in free atoms of its own kind. Once immersed the structure would start to develop irregular surfaces that would then cause it to become more prone to attracting more molecules. This causes the structure to morph even further thus attracting more and more molecules. The pattern continues to repeat itself by capturing more free atoms ultimately creating a larger component of the nanosensor. The second method of self-assembly is more difficult. It starts with a complete set of components that automatically assemble themselves into a finished product (the nanosensor). This method thus far has only been accomplished in manufacturing microsized computer chips. If this method were to be perfected at the nanoscale, the sensors would be made accurately, at a quicker rate, and for a cheaper cost. This is because they would be able to assemble themselves without having to assemble each sensor individually (Whitesides, 1995).

4.3 **TYPES OF NANOSENSORS**

4.3.1 **CHEMICAL NANOSENSORS**

A chemical sensor uses capacitive readout cantilevers and electronics to analyze a transmitted signal. This sensor is sensitive enough to detect a single chemical or biological molecule. Generally, chemical sensors are used to detect very small amounts of chemical vapors. Different types of detection elements, such as carbon nanotubes, zinc oxide nanowires, or palladium nanoparticles can be used as chemical sensors (UnderstandingNano, 2007). These detection elements change their electrical characteristics, such as resistance or capacitance once they absorb a gas molecule. Due to the small size of the detection elements, only a few gas molecules are sufficient to change the electrical properties of the sensing elements allowing for high sensitivity and selectivity. An image of a nanotube-based chemical sensor is shown in Fig. 4.2. The conducting properties of the nanotube change when chemicals in the surrounding environment bond to the tube. The absorbed molecules can act as dopants, shifting the energy of the nanotube. Similarly, the bonds formed between absorbed chemicals and the nanotube change the band structure of the tube (Peng et al., 2001).

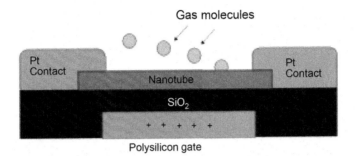

FIGURE 4.2

Schematic of a nanotube-based chemical sensor.

From Peng, S., O'Keeffe, J., Cho, C. W., Kong, J., Chen, R., & Dai, N. F. (2001). Carbon nanotube chemical and mechanical sensors. Conference paper for the 3rd international workshop on structural health monitoring, 1–8.

4.3.2 NANOBIOSENSORS

The field of biosensors is one of the largest funded areas of research pertaining to nanosensors. This is due to all the potential applications that this technology could lead to, such as early cancer detection, detection of various types of diseases, or even detection of specific types of DNA.

The main components of a biosensor consist of a biological element, a physiochemical transducer, and a detector. The biological element is used to bind the target molecule and must be highly specific, stable under storage conditions, and immobilized (Huefner, 2006). Examples of common biological elements include microorganisms, tissues, enzymes, and antibodies. The physiochemical transducer acts as an interface measuring the physical change that occurs within the reaction at the bioreceptor and transforms that energy into a measurable electrical output. The final detector component receives a signal from the transducer that is passed to a microprocessor where they are amplified and analyzed. The data is then converted to measurable units and transferred to a display or data storage device. Fig. 4.3 illustrates the different components of biosensors as described above.

The nanobiosensor can work on several principles of detection. Physical changes can be of piezoelectric (measuring the differences in mass), electrochemical (measuring the differences in electrical distribution), optical (measuring differences in light intensity), or calorimetric (measure the differences in heat).

4.3.3 NANOSCALE ELECTROMETERS

A nanoscale electrometer is a nanometer-scale mechanical electrometer that consists of a torsional mechanical resonator, a detection electrode, and a gate electrode which is used to couple charge to the mechanical element. An example of an electrometer sensor is shown in Fig. 4.4. An external parallel magnetic field is employed for readout.

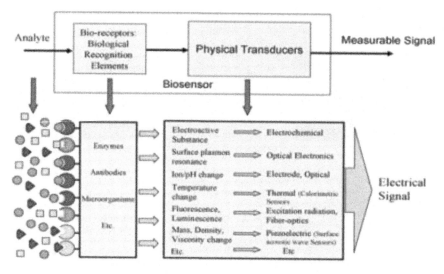

FIGURE 4.3

Biosensor components.

From Maki K. Habib, Biosensors and Bioelectronics 23 (2007) 1–18.

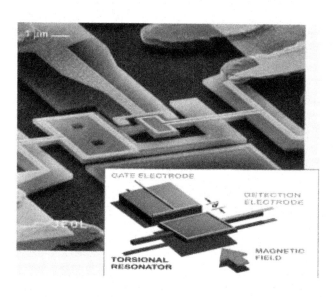

FIGURE 4.4

Principal components of nanoelectrometer.

From Cleland, A. N., & Roukes, M. L. (1998). A nanometre-scale mechanical electrometer. 392.

4.3.4 **DEPLOYABLE NANOSENSORS**

Although deployable nanosensors are not largely funded, they have great uses in military or national security measures. One type of nanosensor that is currently used is the SnifferSTAR, which is a nano-enabled chemical sensor that is integrated into a micro unmanned aerial vehicle (Heil, 2007). This sensor is a lightweight portable chemical detection system that combines a nanomaterial for sample collection and a concentration with a microelectromechanical based chemical lab-on-a-chip detector. This sensor is primarily used in detection of airborne chemical agents and thus could save countless lives from chemical warfare.

4.3.5 **MULTIANALYTE SENSOR ARRAYS**

Most traditional chemical sensors are designed to detect a single chemical species. Nanosensor arrays improve on this not only by detecting the target chemical, but also distinguishing between multiple chemical species in a sample stream. This technology has allowed for vast improvements of real-time chemical monitoring and disease identification (Shelley, 2008). Because nanosensors are so small, an array of sensors—potentially hundreds or thousands—could all be used within a single device.

Sensor chips that carry many different types of nanosensors, each tuned for a specific chemical or biological species, are currently being developed. These sensor chips are essentially a mix of nanowires and nanotubes with different coatings or functional groups. When a nanosensor array is exposed to a sample, the device can generate a unique fingerprint based on the responses of the individual nanosensors as shown in Fig. 4.5.

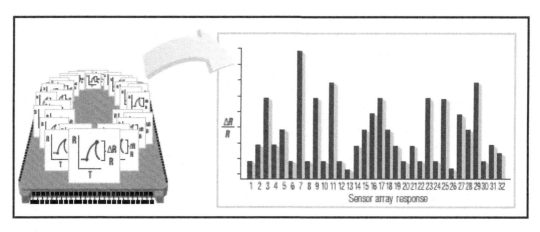

FIGURE 4.5

A device that integrates a vast amount of nanosensor elements.

From Shelley, S. (2008). Nanosensors: evolution, not revolution...yet. CEP, 1–5.

4.4 NANOSTRUCTURES AND MATERIALS

4.4.1 NANOWIRES

Nanowires are small wires at the nanometer scale. They exhibit unique optical and electrical properties at the nanoscale that are not present in the same bulk material. These properties make nanowires a popular nanostructure in the design of different types of nanosensors.

Semiconducting nanowires have electronically switchable properties that provide a label free and direct readout that can be used for sensing. This can be attractive in many applications, because electronic signals from nanosensors can be directly connected and routed to the outside world, which means that they can be easily integrated into miniaturized systems. These properties along with the capability of extreme detection sensitivity make nanowire devices an attractive tool for a wide range of applications.

The main mechanism of nanowire sensor detection is through their configuration as field-effect transistors (FETs). The nanowires exhibit a change in conductivity in response to variations in the potential or electric field at the surface of the FET (Patololsky, 2006).

Typically, a semiconductor (usually Si) is connected to a metal source and drain electrodes through which current is collected and injected. The gate voltage controls the conductance of the FET. In the case of p-Si, applying a positive gate voltage reduces conductance by depleting carriers, while a negative voltage results in accumulation of carriers and increased conductance (Patolsky, 2005). Fig. 4.6 shows a standard FET.

Si nanowires can be fabricated as single crystal structures with diameters as small as 2 nm. They have been shown to exhibit characteristics of performance that are comparable or better than those achieved in the microelectronics industry.

The electronic characteristics of the silicon nanowires are easily controlled during growth, making them easily reproducible. This is another major advantage for this type of nanosensor. Fig. 4.7 shows a Si nanowire array (Greene, 2003).

A general nanowire sensing device as an FET is shown in Fig. 4.8, where binding of a specific protein with a positive charge (red (light shaded grey in print version)) to an antibody receptor (green (dark grey in print version)) yields a decrease in conductance.

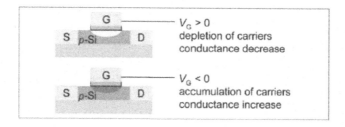

FIGURE 4.6

A standard FET.

From Patolsky, F. L. (2005). Nanowire nanosensors. Materials Today.

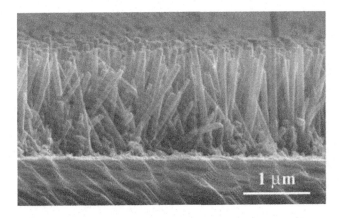

FIGURE 4.7

An array of silicon nanowires.

From Greene, L. L. (2003). Low-temperature production of ZnO nanowire arrays. Chemie International Edition, 3031–3034.

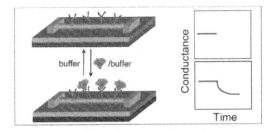

FIGURE 4.8

A general nanowire sensing device.

From Patolsky, F. L. (2005). Nanowire nanosensors. Materials Today.

4.4.2 **CARBON NANOTUBES**

A carbon nanotube is an ordered molecule of pure carbon shaped as a cylindrical structure. Many nanomaterials and nanostructures on their own are poor in terms of selectivity, as they are not naturally selective for anything. For this reason, nanotube-based sensors require some type of chemical engineering. This can be accomplished by adding a coating to modify the surface of the nanotube or by doping atoms into the nanotubes increasing their selectivity to a specific analyte (Shelley, 2008) (Fig. 4.9).

A major advantage of nanotubes is their very high surface area. A single-walled carbon nanotube has a surface area of $1600\,m^2/g$. In addition to their massive surface area, nanotubes are a promising material due to other remarkable properties. They are extremely strong, with strengths ranging from 20 to 100 times that of high strength alloys and steels. Nanotubes are also very resilient, with the ability to

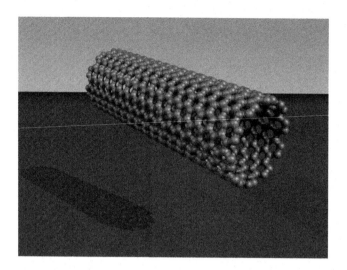

FIGURE 4.9

The structure of carbon nanotubes.

From Wikipedia. (n.d.). Wikipedia. Retrieved 05 12, 2016, from Carbon nanotube: https://en.wikipedia.org/wiki/Carbon_nanotube.

be fully straightened without damage after extreme bending. Other impressive properties include their high tensile strength (about 200 GPa) and their stiffness (roughly five times that of steel). Nanotubes are also ultralight in weight (Shelley, 2008).

Combining this unique set of properties along with the ability to be modified chemically or biologically in order to optimize targeting, nanotubes are the most likely to make their way to widespread commercial applications (Sinha, 2005). Electrical resistance also changes dramatically upon absorption of certain gaseous molecules (e.g., NO_2, NH_3, H_2), and this can be used in chemical sensing. Fig. 4.10 shows a carbon nanotube set up as an FET, where when the analyte of interest is adsorbed, the electrical characteristics of the FET change and can be measured (Zhao, 2002).

Nanotubes can also be used as sensors through the use of mechanical resonance frequency shift to detect adsorbed molecules (Riu, 2005).

The nanotube is free to vibrate on one end (as seen in Fig. 4.11), while the other is attached to a negatively charged electrode. Electrons are able to flow to a positively charged electrode near the free end of the nanotube, and the current of the electrons indicates the frequency at which the nanotube is vibrating. By measuring the current, the frequency is being measured, and this indicates whether or not the particles of interest have become attached or not.

The addition of an adsorbed atom or molecule changes the mass of a vibrating nanotube, and this changes the frequency at which the nanotube oscillates.

4.4.3 THIN FILMS

Thin film nanosensors consist of a thin nanocrystalline or nanoporous sensing film, which is capable of interacting with the environment around it.

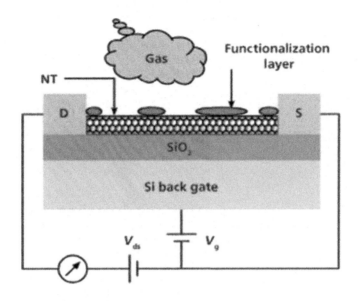

FIGURE 4.10

A nanotube used in an FET device.

From Namomix, S. (2007). A Nanotechnology Test System. EE-Evaluation Engineering, June 2007.

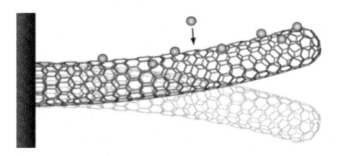

FIGURE 4.11

A vibrating nanotube. As particles attach to the nanotube, the vibration frequency changes.

From Johnston, H. (2008, July 21). Nanotube cantilever weigh up. Retrieved 10.04.11, from Physicsworld: http://physicsworld.com/cws/article/news/2008/jul/21/nanotube-cantilever-weighs-up.

Electrical conductance is measured as it changes when gases adsorb or react at the surface of the nanofilms. Various materials have been used in nanoscaled thin film sensing, such as gold, platinum, diamond, titanium, iridium oxide, and various polymers.

An example of a thin film nanosensor is shown in Fig. 4.12. This nanoporous thin film sensor uses a thiol monolayer that selectively binds to lead or other heavy metals. The film acts as an electrode in the detection of heavy metals. This technology could be used in water sampling (Shelley, 2008).

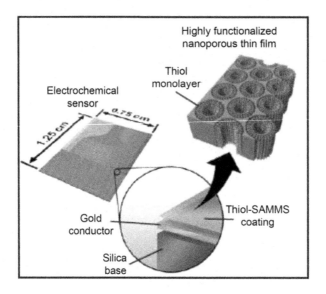

FIGURE 4.12

A thin film nanoscale lead sensor.

From Shelley, S. (2008). Nanosensors: evolution, not revolution...yet. CEP, 1–5.

4.4.4 NANOPARTICLES

Nanoparticles are small clusters of hundreds or thousands of atoms, measuring only a few nanometers in length. The energy level can be tuned and manipulated by synthesizing nanoparticles of different diameters (Riu, 2005). A location that can contain a single electric charge is called a quantum dot (QD). The absence or presence of an electron in a QD changes its properties, and this change in properties can be used for purposes such as information storage or as sensors (Likharev, 1999).

Nanoparticles can be used as nanosensors because they have exceptional size-dependent optical properties. Two types of nanoparticle-based sensors examined in this report are (1) noble metal nanoparticles and (2) semiconductor QDs.

4.4.4.1 Noble metal nanoparticles

Nanostructures of noble metals that are smaller than the de Broglie wavelength for electrons are capable of intense absorption in visible and UV region that is absent in the spectrum of the same bulk material (Riu, 2005).

The de Broglie Wavelength is given by

$$\lambda = \frac{h}{p},$$

where h is Plank's constant and p is the momentum (Richards, 2009).

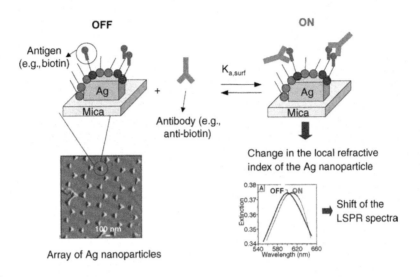

FIGURE 4.13

The biosensing mechanism of a silver nanoparticle using LSPR.

From Riboh, J. H. (2003). A nanoscale optical biosensor: real-time immunoassay in physiological buffer enabled by improved nanoparticle adhesion. The Journal of Physical Chemistry B, 107, 1772–1780.

The excitation of the surface by light is called localized surface plasmon resonance (LSPR). Surface plasmons are electron oscillations at the interface between two materials.

LSPR sensing is based on the sensitivity of plasmon absorbance of metal nanostructures as a response to the changes in dielectric properties of the contacting medium. Typically, a recognition interface is constructed on a metal nanostructure, and specific binding is transduced into an optical signal (Vaskecich, 2008).

The size, material, and shape of the nanoparticle as well as the external environment control the LSPR spectrum. This makes noble metals extremely valuable as possible materials for nanosensors. When target molecules attach to the metal nanoparticles, the local refractive index changes. LSPR spectra are extremely sensitive to these changes in refractive index, and the shift produced in the LSPR spectra can be used to detect molecules. The sensing principle of LSPR sensors is shown in Fig. 4.13. In this example, a silver nanoparticle is used. When the antibody is present, it binds to the antigen, and a change in local refractive index occurs. This results in a shift in the LSPR spectra that can be interpreted as an "on" (presence) or "off" (absence) position.

4.4.4.2 Quantum dots

QDs are small semiconductor crystals that range from a few to a few hundred nanometers. They are commonly made of cadmium serenade, cadmium sulfide, or cadmium telluride. Because of the toxicity of cadmium, QDs usually have an inert polymer coating to protect human cells. The polymer coating is also advantageous in that it facilitates the attachment of various other molecules that can be used to optimize the selectivity of the target species. The conductivity of QDs as well as the wavelength of emitted light can also be manipulated by doping the dots with certain atoms.

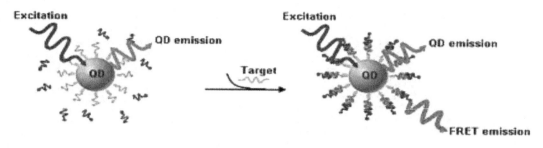

FIGURE 4.14

QD in an FRET reaction.

From Zhang, C.-Y., Yeh, H.-C., Kuroki, M.T., Wang, T.-H. (2005). Single-quantum-dot-based DNA nanosensor.
Nature Materials 4, 826–831.

QDs are able to demonstrate different optical properties as a result of shape, size, and composition, and this can be exploited to custom design optical responses. By controlling the size, shape, or composition of the QD, one can custom design it to emit a certain light under certain conditions, creating a color-coded response.

Semiconductor QDs can be used to develop optical sensors based on fluorescence measurements. A smaller nanocrystal has a larger difference between energy levels and a shorter wavelength of fluorescence. By adjusting the size of the semiconductor QD, any fluorescence colors in the visible spectrum can be obtained. For example, CdSe nanocrystals of about 2.5 nm diameters have green fluorescence, while those with diameters of about 7 nm have red fluorescence (Smith, 2010).

QDs mainly work as sensors because of the Forster resonance energy transfer (FRET) effect, which changes the fluorescence in nanoparticles into either an "on" or "off" state.

FRET is a nonradiative energy transfer from a donor fluorophore to an acceptor fluorophore in close proximity. When a donor fluorophore is in an excited state due to an outside energy source, energy from a donor can be transferred to an acceptor fluorophore located within a range of about 2–8 nm. This transfer of energy will only occur if there is sufficient spectral overlap between the donor emission and the acceptor absorption. Upon energy transfer, the donor fluorescence is reduced, and acceptor fluorescence is increased (Barroso, 2011).

QDs have several characteristics, such as having broad excitation spectra making them ideal candidates for donor fluorophores in an FRET reaction (Fig. 4.14).

4.4.5 POROUS SILICON

Porous silicon is identical to the silicon used in many technological applications today with the difference that the surface contains tiny pores, ranging from less than 2 nm to microns, capable of absorbing and emitting light. The material is a complicated network of silicon threads of thicknesses that range from 2 to 5 nm. This semiconductor material has a very large internal surface area to volume ratio, which can be as high as 500 m^2/cm^3. Light emission takes place in the visible region of the electromagnetic spectrum and is due to quantum confinement effects. The unique property of this material is that the porosity of the material controls the wavelength of the emitted light. With a lower material porosity, longer wavelengths will be emitted, and with a higher material porosity, shorter wavelengths will be emitted. For example, materials with about 40% porosity will emit red light, while materials with a porosity of greater than 70% emit a blue/green light (Riu, 2005).

The luminescence porous silicon can also be altered when molecules are incorporated into the porous layer. This unique property has allowed for the design of gas sensors in which the response can be monitored by visually observing a change in colors.

4.5 APPLICATIONS

4.5.1 MEDICINE

The world of medicine and medical diagnostics is a rapidly changing environment that is hard pressed to keep up with technological advancements in the nanotechnology field. Nanotechnology has enabled the use of nanomedicine in the medical industry by providing a foundation for the development of nanomaterials, nanoelectronic biosensors, and even possible applications of molecular nanotechnology in the near future. Nanosensors are particularly interesting to the medical field because they enable early detection of various diseases along with providing a solid medical diagnostic to the afflicted patient. Early breath detection is a developing nanosensor field where the human breath is analyzed to detect different types of gases that have been shown to announce the presence of certain types of diseases (Gouma, 2010; Stanacevic, 2004). Nanosensors also have the uncanny ability to detect early stages of cancer, which would increase the chance of patient survival (Peng, 2010; Tisch, 2010). The use of nanosensors to detect mitosis and cell atrophy has the potential effect for cell immortality. Nanosensors have the ability to increase the comfort of diabetic lives through the use of a smart tattoo (Samuel, 2011). The nanosensors present in the tattoo will allow for the continuous detection of glucose, ultimately increasing the living capabilities (Kevin, 2010).

The human breath is a complex composition of various gases. It is composed of over 400 different types of gases and molecules with 30 of those already identified as biomarkers for different diseases (Gouma, 2010). The introduction of nanosensors and their specific ability to detect concentrations in the parts per billion scales of gases such as nitric oxide, acetone, ammonia, and carbon oxide has allowed for early disease detection (Gouma, 2010) (Stanacevic, 2004). The presence of nitric oxide, for example, constitutes with the development of asthma attacks. The presence of ammonia can tell whether or not a lung infection is viral or bacterial. The presence of carbon oxide in the human breath can tell whether or not a cardiovascular disease is prevalent in the subject (Stanacevic, 2004). These nanosensors that are developed for this purpose are usually handheld devices that are easily transportable. They usually work by detecting the electrical current that is produced by the dipole gases and translating that measurement into resistance. The resistance then produced is correlated with the amount of the gas that is present in the human breath (Gouma, 2010) (Stanacevic, 2004).

Nanosensors have also been shown to be able to detect early stages of tumor growth through gene/protein changes by the emission of volatile organic compounds (VOCs) in patients' breaths (Peng, 2010). This method involves the use of gold nanoparticles (GNPs) that measure the different amounts of VOCs that are present in the patient's breath (Tisch, 2010). The measurement is conducted through the rapid electrical resistance change within the nanosensor when it is exposed to the VOC. These nanosensors are able to detect different types of cancers such as lung, colon, breast, and prostate and allow for early detection, thus ultimately increasing the chance of patient survival (Peng, 2010; Tisch, 2010).

The study of human telomerase in the human body is an interesting topic because of its ability to lead to cell immortality. Human telomerase is a ribonucleoprotein reverse transcriptase, an enzyme that catalyzes the addition of telomeric ends of each chromosome. When this telomeric sequence reaches a critical length, the cell enters cell atrophy and perishes. Human telomerase is extremely hard to detect, and therefore developing a nanosensor that is able to detect its enzymatic activity and presence is a step in the right direction. The study of human telomerase in cancer cells (which contain an abundant amount of telomerase due to their inability to stop replication) will enable scientists to see how cells are ultimately reproduced (Perez, 2008).

In recent years, there has been a shift in focus to develop a nanosensor that is capable of directly measuring glucose levels. This shift has brought about a fluorescent-based nanosensor. These nanosensors offer the benefit of providing continuous monitoring and can be implanted into the skin, an approach that is referred to as the smart tattoo. This allows the tattoo to change its fluorescent properties in response to the blood glucose level of the diabetic patient. If the glucose level falls, the tattoo properties will change and tell the patient that insulin is needed within the blood (Kevin, 2010; Samuel, 2011).

The use of nanosensors in the medical field for early detection of various diseases through human breath will have a profound effect on medical diagnostics in the future. This will allow doctors to monitor for certain biomarkers allowing them to diagnosis the onset of diseases and cancer in patients. The ability to do this accurately and quickly will increase the longevity of patient life. The ability to create a smart tattoo that detects glucose levels in blood will vastly increase the comfort of diabetic patients. The smart tattoo will allow for continuous monitoring and allow patients to forgo archaic glucose measurement techniques. The study of human telomerase and how it affects cell reproduction will allow scientists to see how tumors ultimately function, thus increasing the chance of a cure for cancer. All of this is possible with the advent of nanosensors and their ability to heavily influence the medical field.

4.5.2 SECURITY

The ability for nanosensors to detect harmful chemical and biological constituents in the atmosphere is becoming a largely funded field. This area of research is of keen importance to various national defense companies and governments. The ability to detect trace amounts of explosives such as trinitrotoluene (TNT) by the use of nanosensors is a prime example of how the nanosensors can be used for preventative security measures. Nanosensors that are able to detect chemical, biological, nuclear, and radiological hazards are also currently in development today.

Nanosensors that are capable of detecting trace amounts of various explosives, such as TNT, in the presence of interferents, such as air or other gases, are in the early stages of development. The sensor that is able to detect for TNT currently employs a hybrid detection mechanism that simultaneously measures the electrochemical current from the reduction of TNT and the change in conductance associated with that reduction. This sensor has useful applications in civilian and military industry. For example, it can be used to detect for explosives on boarding passengers or to detect for explosives in a live battlefield (Aguilar, 2010).

Nanosenors are also able to detect biological and chemical containments that can be used as weapons. Nanosensors that can detect for these constituents use fluorescent nanoparticles, also known as QDs, which are conjugated to fragmented antibodies of the targeted contaminant. When the wavelength dispersive detectors (WDS) are coupled with a quencher (i.e., an antigen), it allows for sufficient fluorescent resonance energy transfer to quench the QD emissions. A subsequent addition of targeted antigen then displaces the bacteria eliminating the resonance energy and causing an increase in QD photoluminescence. This novelty allows for a large application of broad range detection for bacterial and chemical containments. This will allow for quick and easy methods to be developed that will test for any time that biological or chemical agents have already been used. This early detection is key since the proper countermeasure agent can then be administered to the group of people that have been afflicted with the chemical or biological weapon (Ashok, 2009).

These are two quick examples of how nanosensors can help in the security industry by allowing for detection of explosives and or different types of weapons. This knowledge can be used to accurately and quickly treat any threats or diseases that are present in the vicinity.

4.5.3 ENVIRONMENTAL

Nanosensors also have various applications in the environmental field. The ability to sense for chemicals and biological agents that are present in the air and water is a concern to environmental agencies. Nanosensors will innovate the ways air and water quality is measured due to their size, quickness, and accuracy of measurements. An example of this is detecting mercury in any medium (such as air and water) through the use of dandelion-like Au/polyaniline (PANI) nanoparticles in conjunction with surface-enhanced Raman spectroscopy (SERS) nanosensors (Wang et al., 2011).

The ability for nanosensors to measure air quality, particularly for pollutants, is a new approach to air sampling. Nanosensors have already been used to measure solar irradiance, aerosol cloud interactions, climate forcing, and other biogeochemical cycles of East Asia and the Pacific region. Such instrumentation has been useful in tracking air pollution in Beijing during the summer Olympic Games (Dybas, 2008). Nanosensors have also been used by an Israeli start-up company that will monitor and analyze emissions from vehicle engines in order to meet the ever increasing strict standards of American and European Environmental agencies (Brinn, 2006).

Nanosensors can also be used in monitoring water distribution and water quality. Due to the loss of water from leaky pipes and mains, the Environmental Protection Agency has designed an innovative way to improve the water supply infrastructure via a highly cost-effective monitoring system. This "Smart Pipe" prototype is built from a multisensor array that will monitor water flow and quality using nanosensors. It will allow for real-time monitoring of flow rates, pipe pressures, stagnant points, slow flow sections, pipe leakage, backflow, and water quality without altering flow conditions in the already-existing infrastructure (Lin, 2009).

These are just some applications of nanosensors that are used in the environmental field. There are numerous others applications in sensing environmental disturbances, but the two most popular ones are: air quality and water quality/quantity measurements.

4.5.4 **INDUSTRIAL**

Nanosensors also have useful applications in the industrial field. One of the most prominent applications is the ability to detect various industrial gas leaks. Techniques to detect gas leaks include the use of detecting chemical agents such as differential absorption, light detection, and ranging, along with terahertz frequency-based sensing systems. These two systems are capable of detecting gas at ppm to ppb ranges and are portable and easy to use. The sensor is based on a ridged wave guide with various dielectric materials that are structured periodically in an array. The sensor functions by detecting the concentration of industrial gas changes based on the changes in the effective refractive index of the core in the sensor while the industrial gas fills up the receptor space. This type of sensor is small and portable allowing for easy use and a wide range of applications in detecting industrial gases. The three gases that were tested in this cause were: hydrogen sulfide, carbon dioxide, and sulfur dioxide (Sengupta, 2009).

4.6 **CURRENT RESEARCH AND FUTURE PERSPECTIVES**

As discussed in "Applications" section of this chapter, nanosensor research is heavily funded for its many potential uses. Much work is done to study certain characteristics that make these sensors more effective, such as size or morphology. For example, one study, conducted by Wang et al. (2011), focuses on the assembly of dandelion-like Au/PANI nanocomposites and their application as SERS sensors. The Raman signals of PANI molecules in the nanocomposites were enhanced by adding Hg^{2+} ions to the Au/PANI solution, and therefore directly utilized them as excellent SERS nanosensors for the detection of mercury. Their investigations showed that the nanocomposites consisted of dandelion-like spheres, and that this morphology leads to an ultrasensitive response in comparison to other morphologies. The prepared nanocomposites were successfully used as sensors for the detection of Hg^{2+} ions. Similar nanosensors may be used to detect specific ions in many different contexts, such as the environment or certain solutions.

TNT is an explosive that may be detected with the help of nanosensors. Aguilar (2010) describes a nanosensor that is capable of detecting trace amounts of various explosives, including TNT, in the presence of interferents. This has been a difficult challenge to date. The sensor uses a hybrid detection method by simultaneously measuring the electrochemical current from the reduction of TNT and the change in conductance caused by the reduction. The TNT sensor is based on an integrated sensor chip consisting of electrodes for electrochemical detection and conducting nanojunctions for conduction measurements (Fig. 4.15). The polymer (PEDOT) nanojunction is between WE1 and WE2, and the electrode is at WE3. The Si chip is covered with a thin layer ionic liquid. Heating the TNT particulates to 60°C generates TNT vapor which is collected by the ionic liquid layer. The analyte is then reduced and electrochemically detected by the electrode (WE3) and the reduction products are detected by the PEDOT nanojunctions.

This nanosensor can detect quantities as low as 30 pM (or 6 ppt). This detection limit is significantly better than existing electrochemical methods. Furthermore, it has the ability to discriminate analytes from common interferences. Fig. 4.16 shows the graphs containing the results after testing the sensor in a variety of environments. As can be seen, the sensor is successful, as the response is significantly stronger to TNT than to the interferents. Even in the presence of various interferents, it is capable of detecting TNT at the parts-per-trillion level within a few minutes.

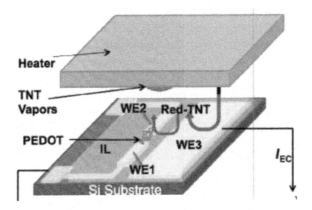

FIGURE 4.15

A hybrid nanosensor: this sensor was used to detect for TNT. See the text for more information.

From Aguilar, A. (2010). A hybrid nanosensor for TNT vapour detection. Nano Letters. American Chemical Society, 380–384.

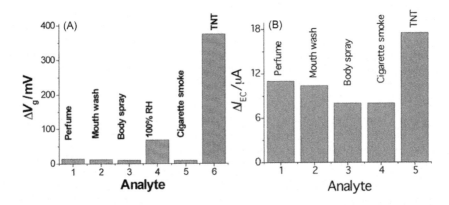

FIGURE 4.16

Sensor response to different analytes. (A) ΔV_g is the shift of the I_d versus V_g curve of the PEDOT nanojunction. (B) ΔI_{EC} is the electrochemical reduction current measured by WE3 at -1.2 V. The response to TNT is stronger in both graphs.

From Aguilar, A. (2010). A hybrid nanosensor for TNT vapour detection. Nano Letters. American Chemical Society, 380–384.

A promising application for nanosensors is their use for breath analysis. Analysis of bodily fluids (blood, sputum, and urine) is important for disease diagnosis and monitoring. However, noninvasive ways, such as breath analysis, are currently underdeveloped. Among 400 compounds that are in human breath, only 30 have been identified and most are potential indicators for more than one disease. For example, CO levels are markers for cardiovascular diseases, diabetes, nephritis, etc. Gouma et al. have developed a metal oxide-based nanosensor to detect ammonia gas in a breath-simulating environment at ppb concentrations, with a lower bound of 50 ppb. Ammonia is important because it can differentiate between viral and bacterial infections in lung diseases to justify the use of antibiotic. The sensor was shown to be reversible; response and recovery times were extremely fast (milliseconds to seconds), and the expected lifetime is over 1 year at operation temperature (Gouma, 2010). Furthermore, the nanosensor was shown to work up to 25% humidity with no effect on performance.

Many other nanosensors have been developed to detect certain substances in the breath. Yet another example is a nanosensor prototype by Wang et al. (2011) to detect acetone in a single breath sample. Acetone is a biomarker for type 1 diabetes. This specific sensor's detection mechanism is done by ε-WO_3, a type of ferroelectric material that has spontaneous electric dipole moment. This polarity comes from the displacement of the tungsten atoms from the center of each WO_3. Acetone has a much larger dipole movement than any other gas, and as a consequence, interaction between the acetone and WO_3 produces a strong current that is detectable.

Breath analysis can also be used to detect certain types of cancers. Tumor growth involves gene/protein changes which may lead to oxidation of the cell membrane. This can be detected by the emission of VOCs in the patient's breath. Peng (2010) investigated the ability to distinguish between breath VOCs that characterize healthy states and those that indicate different types of cancer using a nanosensor array. Breath samples were collected from 177 individuals aged 20–75 years of age. The volunteers included several different types of cancer patients (lung, colon, breast, and prostate), as well as healthy controls. A nanosensor array made of 14 GNP sensors was used. Each sensor goes through a rapid change in electrical resistance when exposed to the sample. These resistance changes were measured and the signals were analyzed using principal component and cluster analysis. The results showed that the nanosensor array was capable of differentiating between the breath of cancer patients and the healthy controls as shown in Fig. 4.17A–D. It was also capable of distinguishing between the different cancer types (Fig. 4.17E). Such nanosensors could lead to the development of a noninvasive, easy to use, inexpensive alternative method for cancer diagnostics.

Cancer can also be detected by analyzing the activity of telomerase. Human telomerase is a ribonucleoprotein reverse transcriptase that catalyzes the addition of telomeric units to the telomere ends of each chromosome (Perez, 2008). It is a key oncogenic gene. Currently there are a number of novel telomerase inhibitors that are being developed. Current technologies used to measure the presence and enzyme activity are time consuming and prone to false positives or false negatives. Perez (2008) has developed a set of function magnetic nanosensors capable of measuring the concentration and enzymatic activity of telomerase. The method of detection is based on a magnetic relaxation switch assay that is able to detect the presence of telomerase protein activity and presence in various cancer and normal cell lines. Two sets of nanosensors are involved. One set of magnetic nanoparticles is conjugated to telomeric repeat oligonucleotides, resulting in nanosensors able to measure telomerase activity. Another set is conjugated to anti-hTERT antibody, resulting in a nanosensor that detects different amounts of telomerase protein. Measurements of telomerase activity and presence are performed by measuring the difference of T2 in a water relaxation environment (Grimm, 2003) (Fig. 4.18).

Nanosensors may also be used to detect health conditions other than cancer. One example is diabetes. Currently there is no cure for diabetes. At the individual level, diabetes is managed by the monitoring and control of glucose levels in the blood in order to minimize the effects of the disease. Today, diabetes patients typically do this by obtaining a small sample of blood, usually a finger prick, which is then placed on a sensor strip and read by an electronic reader which reports the blood glucose concentration. This method has limitations such as painful sampling and noncontinuous monitoring.

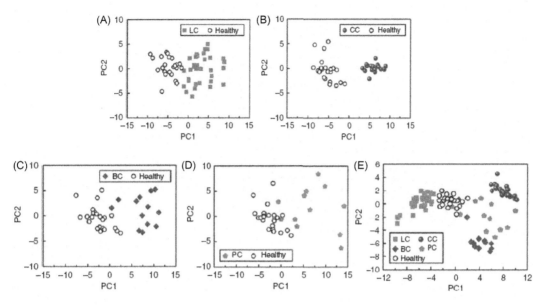

FIGURE 4.17

Results from the cancer detection study. PCA plots of the GNP sensor array's resistance responses of cancer with healthy controls: (A) lung cancer, LC, (B) colon cancer, CC, (C) breast cancer, BC, (D) prostate cancer, PC, and (E) all cancer patients and healthy controls together.

From Peng, G. (2010). Detection of lung, breast, colorectal, and prostate cancers from exhaled breath using a sing array of nanosensors. British Journal of Cancer, 542–551.

Nanomaterials have already improved the size, lifetime, accuracy, and usability of sensors that are currently in use, but in recent years the focus has been on nanosensors that can directly measure glucose levels, and a system that is capable of continuous monitoring.

A solution that is currently being developed involves fluorescent-based nanosensors. These types of nanosensors offer the benefit of providing continuous monitoring and can be implanted into the skin. This approach is referred to as a "smart tattoo" (Fig. 4.19A), where the nanosensors would change fluorescence properties in response to the blood glucose level.

This method could eliminate the need for patients to take blood samples and minimize the risk of infection. Several nanosensors have been developed using fluorescence signals that could possibly be used in this process. One type of nanosensor that could be used as a step toward developing a "smart tattoo," are fluorescent nanosensors that are based on highly plasticized hydrophobic polymers (Fig. 4.19B). The core of the nanosensor consists of hydrophobic boronic acid that is capable of extracting glucose. In the absence of glucose, the boronic acid binds the nonfluorescent alizarin, while in the presence of glucose, the boronic acid binds glucose, releasing alizarin and decreasing overall fluorescence. The sensors are reversible because the components are hydrophobic and remain in the core of the sensor. This allows for continuous monitoring. The development of these nanosensors represents a major step in the development of a smart tattoo for continuous, real-time glucose monitoring in diabetes patients.

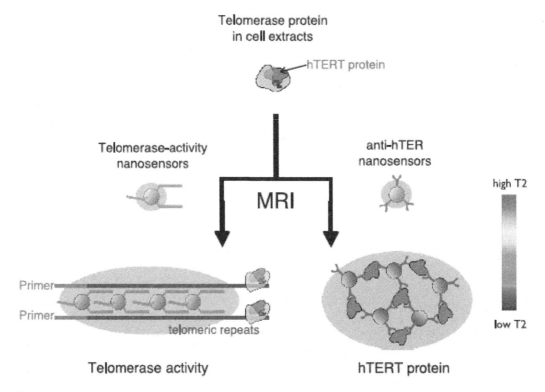

Telomerase protein
in cell extracts

hTERT protein

Telomerase-activity
nanosensors

anti-hTER
nanosensors

MRI

high T2

low T2

Primer

Primer

telomeric repeats

Telomerase activity

hTERT protein

FIGURE 4.18

Process of using MRI and nanosensors for enzyme and human telomerase detection. The cell extract containing telomerase protein is incubated with a sensor in solution. T2 relaxation time changes (induced by clustering of nanoparticles; blue (dark grey in print version)) are measured, as they are proportional to the levels of telomerase activity and amount of telomerase protein.

From Perez, J. M. (2008). Integrated nanosensors to determine levels and functional activity of human telomerase. Neoplasia.

Nanosensors can be of much importance in cellular biology. For example, Kuang and Walt (2007) used self-assembled fluorescent nanosensors to detect oxygen consumption in the proximity of *Saccharomyces cerevisiae*. Oxygen concentration can provide information about certain cell characteristics, such as their "viability upon exposure to cytotoxic drugs and environmental stress, protein synthesis capability, mitochondria function," etc. The synthesis of the nanosensors began as amine functionalized 100 nm polystyrene nanobeads were first covalently conjugated with polyethylenimine (PEI) and soaked with a ruthenium (II) complex that was oxygen-sensitive. This produced a florescent oxygen nanosensor with ruthenium (II) complex entrapped throughout the inert but highly oxygen-permeable polystyrene matrix. The resulting oxygen nanosensors were then assembled on cell surfaces via electrostatic interactions between the positively charged PEI and the negatively charged *S. cerevisiae* cell surface at a physiological pH of 7.4. Once the oxygen nanosensors were ready, the authors incubated them with cultured yeast cells at different cell/nanosensor ratios. The number of nanosensors on each cell

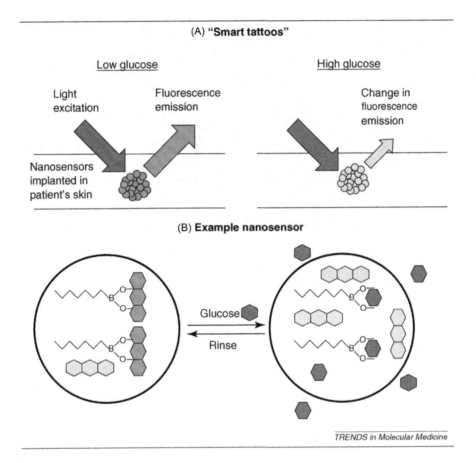

(A) "Smart tattoos"

Low glucose | High glucose

Light excitation | Fluorescence emission | Change in fluorescence emission

Nanosensors implanted in patient's skin

(B) Example nanosensor

Glucose

Rinse

TRENDS in Molecular Medicine

FIGURE 4.19

Nanosensor solution for diabetes. (A) Smart tattoo. (B) A type of nanosensor based on highly plasticized hydrophobic polymers.

From Kevin, C. (2010). Nanosensors and nanomaterials for monitoring glucose in diabetes. Trends in Molecular Medicine. 16.

was controlled by varying the concentration of the stock nanosensors. The results showed that cells covered with higher numbers of nanosensors showed strong fluorescence (Fig. 4.20).

Yeast cells coated with oxygen nanosensors were deposited into a single cell microwell array such that many cells were individually and simultaneously monitored. The authors' results successfully tracked the temporal profile of oxygen concentration change measured by the nanosensors. The authors concluded that this demonstrated that the dynamics of oxygen consumption can be noninvasively and sensitively measured in the proximity of individual cells with self-assembled nanosensors. The nanosensors measured the oxygen concentration directly at the cell surface, suggesting that other analytes with slower diffusion rates and more rapid kinetics could be measured in a similar manner.

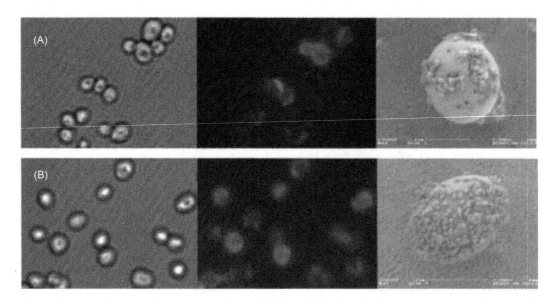

FIGURE 4.20

White light, fluorescence, and SEM images of *S. cerevisiae* cells incubated with (A) 1% and (B) 5% concentration of stock oxygen nanosensors.

From Kuang, Y., & Walt, D. R. (2007). Detecting oxygen consumption in the proximity of Saccharomyces cerevisiae cells using self-assembled fluorescent nanosensors. Biotechnology and Bioengineering, 96(2), 318–325.

Since nanosensors are still in development, much research is also done to investigate methods of assembling these sensors. Many studies have been published examining novel ways of assembling certain types of nanosensors. One example is the work by Blum, Soto, Sapsford, and Wilson (2010), who have examined the bottom–up self-assembly for electronic nanosensors, in which the recognition of a binding event transfers directly into a readable electronic signal. The designed nanosensors consisted of a molecular electronics-based network built on a cowpea mosaic virus (CPMV) capsid scaffold. The CPMV was used for binding cysteine (cys) residues and anchoring GNPs to specific locations on the scaffold. The gaps between the GNPs were adjusted such that a 3D conducting network could be constructed on the nanoscale. The conductance of such molecular networks can be measured upon exposing them to the analyte.

As evident from the wide variety of studies done, there is no doubt that nanosensors have great potential, as they continue to provide more effective and simpler alternatives to many existing sensors.

4.7 CONCLUSIONS

Nanotechnology is in its infancy, and has potential applications in virtually every industry, with new applications certainly to be discovered. Nanosensors, in particular, provide a bridge between two worlds by conveying information from the nanolevel to the macroscopic world, allowing for detection at the parts per billion level.

The major advantage of nanosensors lies in their extremely high surface area to volume ratio, which allows for unprecedented levels of sensitivity and the detection of a single molecule or atom. At the nanoscale, materials have unique optical and electrical properties, require less power to operate, and have greater selectivity. For these reasons, nanosensors have the potential to revolutionize detection methods in medical diagnostics, environmental monitoring, and many other applications.

The major challenge to date is the ability to produce cost-effective, fabrication methods that are reproducible on a large scale. The wide range of potential benefits of using sensors at the nanoscale, along with the massive increase of interest, research, and development of nanotechnology, in the last decade has led to major advancements within the field. The question therefore is not *if*, but *when* nanosensors will be used in a wide range of everyday real-world processes.

NANOPARTICLES[*]

5.1 INTRODUCTION

The terms nanotechnology, nanostructure, nanoscience, and nanoparticles have recently been used in scientific and engineering literatures widely. It was found that nanoscale materials are very applicable and useful for devices that will be able to travel through the human body and repair damaged tissues or supercomputers. It was also found that these nanoscale materials have applications in many other fields such as biological detection, controlled drug delivery, low-threshold laser, optical filters, and also sensors, among others (Fredy, 2008).

The rapid development in the chemistry of nanoparticles over the past 20 years has driven tremendous advances in the field. While there remains significant interest in the use of nanoparticles as fillers in polymer materials to enhance physical and mechanical properties, many are now engaging in research efforts that focus on precise structures of nanoparticles in polymers, including their assembly in arrays and along interfacial boundaries. The combination of precise organic chemistry on nanoparticle materials has led to developments in the growth of polymers from nanoparticle surfaces, allowing one to tailor the properties of the particles by the choice of polymer and its inherent functionality. Nanoparticle–polymer composite materials are generating interest in several applications, including electronic and optical materials, based on the properties of the metallic and semiconductor particles used. In addition, efforts in catalysis are under way using nanoparticles in polymer scaffolds. Nanotechnology is revolutionizing the scientific approach across many disciplines, including chemistry, biology, materials science, engineering, and theory. For example, research centered on nanoscopic materials extends from the semiconductor industry, where the ability to produce nanometer-scale features leads to faster and less expensive transistors, to biotechnology, where a variety of uses are available (Skaff, Emrick, & Rotello, 2008).

5.2 SYNTHESIS AND CHARACTERISTICS

Synthesis of nanoparticles makes it possible to have some degree of control over the nanoparticle shape and stability. The synthesis of these particles requires precautions to avoid their aggregation. It also sometimes involves the use of a stabilizing agent, which associates with the surface of the particle and

[*] By Yaser Dahman, Hoda Javaheri, Jiafu Chen, and Basel Al-Chikh Sulaiman.

Nanotechnology and Functional Materials for Engineers. DOI: http://dx.doi.org/10.1016/B978-0-323-51256-5.00005-8

provides charge or solubility properties to keep the nanoparticles suspended, and thereby prevents their aggregation (Fredy, 2008).

5.2.1 MAGNETIC NANOPARTICLES

5.2.1.1 *Characteristics of magnetic nanoparticles*

Among all kinds of nanomaterials, magnetic nanoparticles earn their position by their special features and wide-used applications. Compared to regular magnetic materials, nanoparticles differ from the domain structure to the classic quantum theory. Thus, they have a more advanced technology and more applications due to different physical and chemical properties.

The physical properties of magnetic nanoparticles can be determined by the chemical compositions, the type and the degree of defectiveness of the domain structure including the particle size and shape, and the interaction of the atoms among the molecular structure. However, due to the limitations of present technology, the above factors cannot be controlled all the time during the synthesis. Furthermore, the relationship between the properties and the structure is unknown, so the same type of material with different concentrations could have big differences.

Even though magnetic nanoparticles are basically metals with magnetism, living creatures also have magnetic nanoparticles within their bodies. For examples, migratory birds and fish have magnetic nanoparticles inside their sense system to guide them during their migration between south and north every year. Even human beings have magnetic nanoparticles inside their brains. It is estimated that the human brain includes about 100 million magnetic nanoparticles per gram of tissue. So magnetic nanoparticles are not just a science, but are related to our daily life and affect it in many different ways.

There are many kinds of magnetic nanoparticles due to different chemical properties: metals, rare earth metals, oxidation of metallic nanoparticles, and magnetic alloys. Since the metal nanoparticles include most of the metal magnetic materials and oxidation of metallic nanoparticles, the following section will be focusing on the different magnetic alloys (Gubin, 2009).

Fe–Co alloys: Since Co and Fe are body-centered cubic structures in a nanoparticle, the allied order of both metals is very soft and very suitable to be raw material for nanoparticles. The maximum concentration of Co in the alloy is 35%. This is the saturation concentration of Co. The related magnetic properties also increase with the mixing level (Gubin, 2009).

Fe–Ni alloys: Samples of Fe and Ni in experiments can have nonmagnetic or magnetic soft ferromagnets. For the alloy compound of iron and nickel, they have a much lower saturation magnetization compared to pure samples of each metal. For example, when we have 37% of Ni, it has a low curie point and an FCC structure. Theoretical calculations estimate a more complicated magnetic structure for these types of alloys due to the different combinations.

Fe–Pt alloys: Because of the wide application on the information recording density of materials, these types of alloys are being studied a lot recently. They have the face-centered tetragonal structure and thus obtain a unique property of recording advantage of large coercively when in room temperature, no matter how small the particles are. Prepared by joint thermolysis in the presence of oleic acid and oleyl amine, Fe–Pt nanoparticles have a narrow size distribution. After further heating, a protective film is formed on the surface of alloys, which remain about the same size.

Co–Pt alloys: In high-density information storage field, nanoparticles of Co–Pt have a lot of advantages due to their form, size, and crystal structure, which makes them chemically stable and magnetically crystalline. One of the many methods is the polyol method that does not use organ metallic

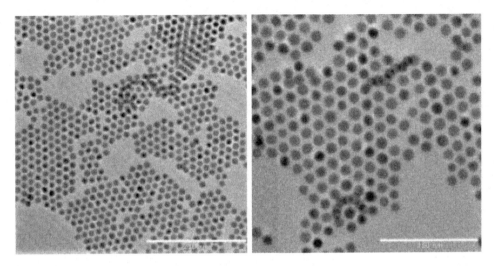

FIGURE 5.1

Transmission electron microscopy (TEM) images at different magnifications of monodisperse Co particles on a carbon coated Cu grid.

From Giersig, M., & Hilgendorff, M. (2005). Magnetic nanoparticle superstructures. Germany: Wiley-VCH Verlag.

precursors. Even though the same solution is processed, sometimes two different concentrations of alloys are formed with different structures, in this case, the different concentration can be studied.

Since most of magnetic materials are metal or metal oxides, the raw material is straight forward and can be easily found. However, the technology difficulties remain in the purity section. As mentioned above, the different chemical properties are based on the structure and the concentration of the alloys, so the synthesis process has to be controlled very carefully. Furthermore, when it comes to the nanoscale, many of the classic physics laws are not applicable and quantum physics is required.

The synthesis of ferrofluid consists of synthetic procedures. Basically, it is the thermal decomposition of metal organic compounds or the thermal decomposition of monometallic metal organic compounds. Fig. 5.1 shows examples of magnetic nanoparticles for one simple structure and for metal alloy compounds.

From Fig. 5.1, we can see the Co particles prepared by thermal decomposition; the raw materials are octacarbonyldicobalt and dichlorobenzene.

Twenty years ago, Massart developed a method to synthesize magnetic ferrofluids based on the co-precipitation of Fe salts in aqueous solutions using repulsive electrostatic forces. The image of Fe oxidation product is slightly different from the Co product, as it is shown in Fig. 5.2.

As we can see from Fig. 5.2, the left side is the iron oxide product prepared in aqueous solution and the right side is the product formed from the toluene solution, which have oleic acid on the surface.

5.2.1.2 Synthesis method of magnetic nanoparticles

The synthesis of magnetic nanoparticles has been developed through over 30 years. The raw materials have been used from metal to nonmetal and from gas to liquid phases. The most commonly used metal

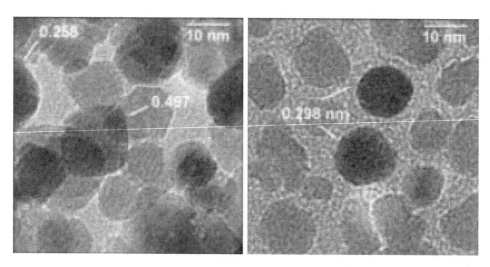

FIGURE 5.2

Typical TEM images of iron oxide particles.

From Giersig, M., & Hilgendorff, M. (2005). Magnetic nanoparticle superstructures. Germany: Wiley-VCH Verlag.

oxides are the Fe, Co, Mg, and Mn with their alloy compounds. In the recent years, many experiments have been done on the control of shape, crystalline, and stable surface of magnetic nanoparticles. As mentioned earlier, the shape and orientation affect the chemical properties greatly. The most common way is co-precipitation, thermal synthesis, and microemulsion. The following section will focus on each of the techniques.

5.2.1.2.1 Coprecipitation

For most of the iron oxides, this is a very convenient way to process aqueous Fe salt solutions. By adding a base on normal temperature and pressure, the process could be controlled to get ideal shape and size of the magnetic nanoparticles. However, there are also other factors to consider with this kind of method: the Fe ratio, the reaction rate and temperature, and the pH value of the solution. These will also affect the smoothness of the reaction. After the preparations are done, the experiment will proceed to a point where the solution reaches magnetic saturation (Lu, Salabas, & Schth, 2007).

5.2.1.2.2 Thermal decomposition

In order to control shape and size more precisely, the method of thermal decomposition is developed. This is a method similar to the synthesis of semiconductors with high-quality nanocrystals. The smaller size magnetic nanocrystals can be formed from organometallic compounds in organic solvents. By adding precursor in zerovalent, thermal decomposition will have metal formed in the end. When the decomposition happens, cationic metal will lead the electrons to the oxides and the reaction solution will have metal acid salts in non-aqueous solution. With the metal fatty acid compound in the solution, the metal will reach to saturation and the metal magnetic nanoparticles will precipitate out from the solutions.

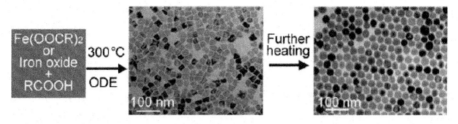

FIGURE 5.3

The process of formation of Fe oxide nanocrystals.

From Lu, A.-H., Salabas, E., & Schth, F. (2007). Magnetic nanoparticles: Synthesis, protection, functionalization, and application. Weinheim: Verlag GmbH & Co. KGaA.

Fig. 5.3 demonstrates the process of thermal decompositions.

Fig. 5.3 shows the TEM images of synthesized nanocrystals at different reaction times. At the beginning, the crystals are in random arranged orders and with random size arrangements. After further heating, the order is clear at the end with regular size and shape.

5.2.1.2.3 Microemulsion

A microemulsion process is the process that two immiscible liquids will go through a dispersion stage where both liquids are stabilized by the surfactant molecules. For example, oil and water are two immiscible liquids. The disperse phase of oil will be surrounded by surfactant molecules. With the desired reactants in the solution, the micro droplets will form and then break to precipitate in the micelles. Acetone and ethanol can be used as solvent for micro emulsions to extract the precipitate by changing the structure. Fig. 5.4 shows the nanorod synthesis process (Lu et al., 2007).

From Fig. 5.4, we can see the different compounds and structures are formed by the different raw materials. The steroid acid products are straight structures, while the octanoic acid products are short and thick.

5.2.2 METAL NANOPARTICLES

5.2.2.1 Synthesis methods

In general, the production of metallic nanoparticles can be categorized into two different kinds: top–down and bottom–up. The first three methods explained below fall under the bottom–up method in which atoms are assembled into nanoparticles, and the last method (4) is the top–down approach in which nanoparticles are cut from larger structure (Blackman & Binns, 2008).

1. *Production of preformed nanoparticles in the gas phase*: Several methods have been found to produce gas-phase nanoparticles. However, they all involve the production of a super-saturated metal vapor that condenses into particles. This method is the most flexible technique for the synthesis of metal nanoparticles, and it is the way to produce tightly mass-selected nanoparticles of virtually any material or alloy in environments ranging from free particles in vacuum to embedded nanoparticles in solid matrices (Blackman & Binns, 2008).

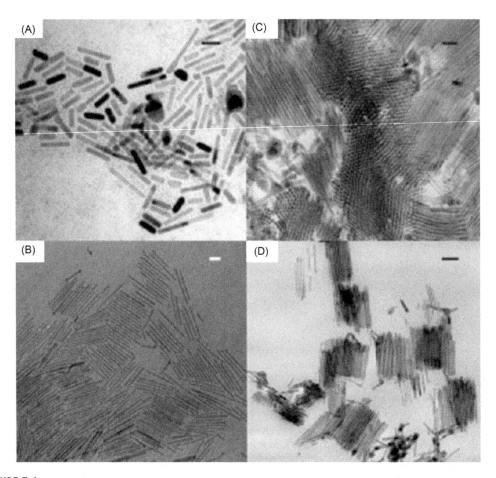

FIGURE 5.4

TEM micrographs of nanorods. (A) Octanoic acid, (B) lauric acid, and (C and D) steroid acid.

From Lu, A.-H., Salabas, E., & Schth, F. (2007). Magnetic nanoparticles: Synthesis,
protection, functionalization, and application. Weinheim: Verlag GmbH & Co. KGaA.

2. *Deposition and self-assembly on the surface*: To explain this technique in the simplest case, take the Volmer–Weber growth on low surface energy materials such as graphite, where a degree of size control can be achieved by optimizing the coverage, substrate temperature, and deposition rate. This can be explained with an example that a method is to deposit onto surfaces with a natural patterning, such as herringbone reconstruction on Au (1 1 1) surfaces on which metal islands tend to nucleate at specific points on the reconstruction. The metal islands formed are in ordered arrays, which is an advantage for some experimental techniques. Electro-deposited films are also included under this generic heading (Blackman & Binns, 2008).

3. *Wet chemical methods*: There are a number of methods to reduce metal salts and obtain aggregation into metallic nanoparticles as a suspension in a solvent. Chemical methods are used to produce a wide range of metal nanoparticles comprising one or more elements. One of the most popular methods is called the polyol method. In this method, monodisperse assemblies of FePt alloy nanoparticles with sizes ranging from 3 to 10 nm are produced via the reduction of $Pt(acac)_2$ (acac: acetylacetonate) combined with the thermal decomposition of $Fe(CO)_5$ in the presence of oleic acid and oleyl amine (Blackman & Binns, 2008).

4. *Top–down methods*: Two main methods in this approach are electron beam lithography (EBL) and focused ion Beam (FIB) milling. In the EBL method, the substrate should first be spun-coated with an electron resist material. The most common material used for this purpose is polymethyl methacrylate (PMMA), which is a positive resist, that is, exposure to the electron beam makes it impervious to a solvent that is then used to dissolve away all unexposed areas. After coating substrate with PMMA, it is exposed to an electron beam focused to a spot almost 5 nm. PMMA can also be used as a negative resist, in which only the areas exposed to the electron beam are dissolved. The patterned resist is then used as a mask through which nanoscale metal islands can be deposited onto a substrate. Then after deposition, the rest of the resist material with its metal layer can be rinsed off in a suitable solvent leaving just the nanoscale metal features on the surface (Blackman & Binns, 2008). In the second method of FIB processing, the metal layer has to be cut into nanoparticles and then be deposited onto a substrate. The unwanted parts are milled away using a beam of Ga^- ions. The Ga^- ion beam is focused on a spot almost 5 nm at the surface (Blackman & Binns, 2008). With both techniques, particles as small as 20 nm are produced. They are unstable and as small as bottom–up approach particles. However, these particles are more flexible and can be made with different shapes (rectangular, circular, triangular, elliptical, etc.).

Fig. 5.5 presents the four generic methods of production of metal nanoparticles.

5.2.2.2 Wet chemical methods

The synthesis of metal nanoparticles requires several steps in liquid phase. In the case of liquid solution, the first step is the formation of metal atoms which can be achieved by the reduction of metal precursor(s) by using chemical reductants in the solution. The formed metal atoms can undergo elementary nucleation with the slow growth process leading to the formation of nanoparticles. This model is called La Mer after the name of the first person to describe it in detail. The concept and apparatus of this model are shown in Fig. 5.6A and B, respectively. It should be noted that a good understanding of this process will help the engineering of the particle size and shape (Blackman, 2008).

5.2.2.2.1 Nucleation

Nucleation takes place because supersaturated solutions are thermodynamically unstable. Therefore, in order for the nucleation process to occur, the solution must be supersaturated generating extremely small sized sol particles (Blackman, 2008). The overall free energy change or ΔG is defined as the sum of the free energy due to the formation of a new volume and the free energy due to a new surface created (Blackman, 2008). For a spherical particle, maximum free energy is the activation energy for nucleation and can be calculated from the following equation:

$$\Delta G = -\frac{4}{V}\pi r^3 k_B T \ln(S) + 4\pi r^2 \gamma$$

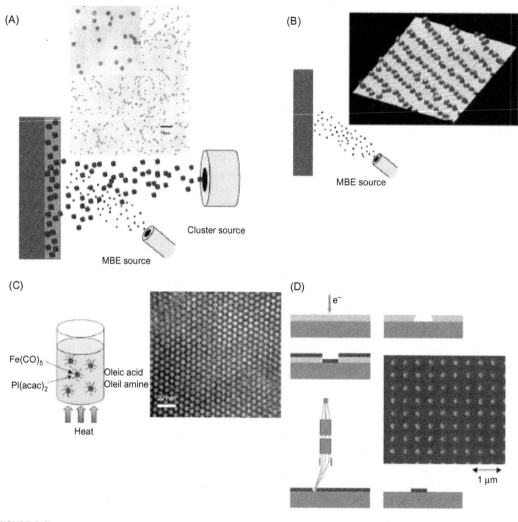

FIGURE 5.5

(A) Production of gas-phase particles, (B) deposition and self-assembly on surfaces, (C) wet chemical methods, and (D) top–down methods.

From Blackman, J., & Binns, C. (2008). Nanoscience and nanotechnology. Handbook of Metal Physics, 5, 1–385.

Where

V is the molecular volume
r is the radius of the nuclei
k_B is the Boltzmann constant
S is the saturation ratio and
γ is surface free energy per unit surface area.

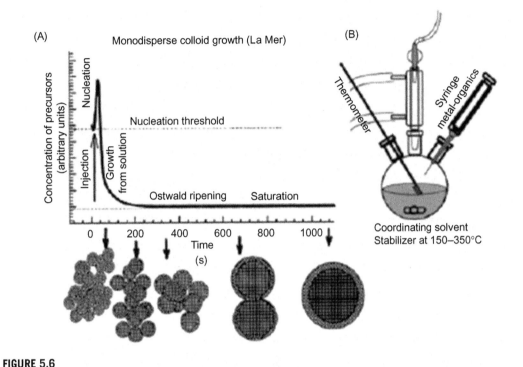

FIGURE 5.6

(A) The concept of monodisperse colloid growth of La Mer model, (B) typically synthetic apparatus.

From Blackman, J. (2008). Metallic nanoparticles. Handbook of Metal Physics, 5, 1–385.

The critical nuclei size, r^*, is obtained from $\mathrm{d}\Delta G/\mathrm{d}r = 0$ (Blackman, 2008):

$$r^* = \frac{2V\gamma}{3k_B T \ln(S)}$$

From this equation, it can be seen that a larger saturation ratio will result in a small critical nuclei size.

5.2.2.2.2 Growth

The next process after nuclei are formed from the solution is the growth process. The growth process takes place via deposition of the soluble species onto the solid surface. If the concentration falls below the critical level, nucleation stops. However, the particles continue to grow by molecular addition until the equilibrium concentration of the species is reached (Blackman, 2008). If both nucleation and growth reactions are stopped rapidly, there is a possibility to obtain a monodisperse size distribution. An alternative strategy for a monodisperse distribution is to maintain a saturated condition throughout the reaction by building up the concentration of metal ions through repeated injections of the metal precursor(s) (Blackman, 2008). An important factor that must be considered in this process

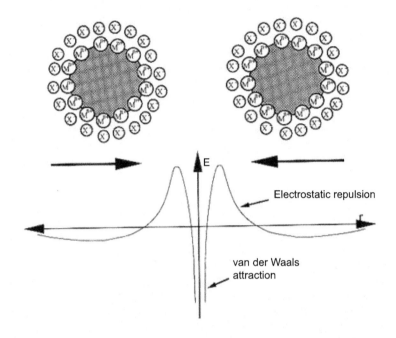

FIGURE 5.7

Electrostatic stabilization of metal colloid particles. Attractive van der Waals forces are outweighed by repulsive electrostatic forces between adsorbed ions and associated counterions at moderate interparticle separation.

From Schmid, G. (2006). Nanoparticles: From theory to application. London: Wiley VCH.

is secondary growth. This is the growth of particles by aggregation with other particles. Growth by this process is faster than that by molecular addition, and it occurs by stable particles combining with smaller unstable nuclei (Yu, Oduro, Tam, & Tsang, 2008). The produced nanoparticles are small and thermodynamically unstable. There are two ways to stabilize these particles, either by adding surface-protecting reagents such as inorganic ligands or inorganic capping materials, or by replacing the particles in an inert environment (Yu et al., 2008).

5.2.2.2.3 Stabilization of colloidal metal particles in liquids

Before describing the synthetic methods, it is important to understand colloidal chemistry, because that is the mean by which the metal particles are stabilized in the dispersing medium, since small metal particles are unstable with respect to agglomeration to the bulk. Two particles at short interparticle distances are attracted to each other by van der Waals forces. In the absence of repulsive forces, to counteract this attraction, an unprotected sol would coagulate. This concentration can be achieved by two methods: electrostatic stabilization and steric stabilization (Schmid, 2006). As an example, classical gold sols are prepared by the reduction of $AuCl_4$ by sodium citrate; the colloidal gold particles are surrounded by an electrical double layer formed by adsorbed citrate and chloride ions and cations which are attracted to them. The resulted Coulombic repulsion between particles is shown in Fig. 5.7.

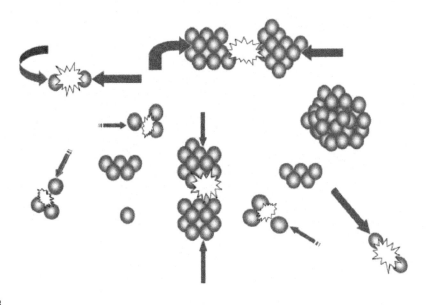

FIGURE 5.8

It is energetically and kinetically favorable for uncharged nanoparticles to collide and coalescence.

From Yu, H., Oduro, W., Tam, K., & Tsang, E. (2008). Chemical methods for preparation of nanoparticles in solution. Handbook of Metal Physics, 5, 1–385.

5.2.2.3 Nanoparticles in aqueous solution

Colloidal nanoparticles in aqueous solution have recently gained much attention for their potential applications. Particle agglomeration may either be required to occur in a controllable manner or needed to be avoided, depending on particular applications (Yu et al., 2008).

In order to understand particle agglomeration, it is important to understand the surface forces at the interface of the particle and aqueous portion of the colloid. Particles with zero charge are induced aggregation; large particles can be obtained by combining them, and they settle by gravitational force. Fig. 5.8 represents nanoparticles with zero charge in liquid phase, in which it is easy for them to collide and form larger clusters.

If particles are carrying like electrical charges, a force of mutual electrostatic repulsion between adjacent particles is produced (Yu et al., 2008). This process is shown in Fig. 5.9.

Encapsulation of the particles by silica in a 7.4 pH buffer is a way to produce stable and homogenous nanoparticles. These particles do not aggregate due to the lower isoelectric point of the silica, rendering the overall particle charge negative. Particles possessing high charges can induce double layers in aqueous environments and are discrete, disperse, and in suspension (Yu et al., 2008). The double layers of colloidal and electrostatic principle are presented in Fig. 5.10.

In contrast with the repulsive energy, there is van der Waals attraction. The combination of an attractive force and a repulsive force induces net interaction energy as shown in Fig. 5.11 (Yu et al., 2008).

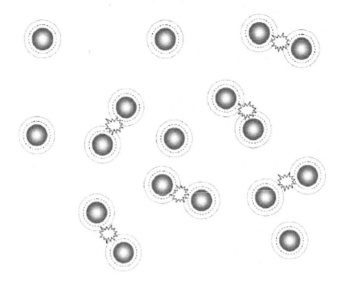

FIGURE 5.9

Double layers of ions can protect colloid particles against rapid aggregation. They can be kinetically stable in an aqueous medium with identical surface charge.

From Yu, H., Oduro, W., Tam, K., & Tsang, E. (2008). Chemical methods for preparation of nanoparticles in solution. Handbook of Metal Physics, 5, 1–385.

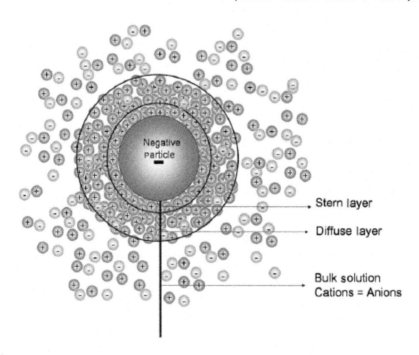

FIGURE 5.10

A colloidal particle protected by a double layer of ions in an aqueous medium.

From Yu, H., Oduro, W., Tam, K., & Tsang, E. (2008). Chemical methods for preparation of nanoparticles in solution. Handbook of Metal Physics, 5, 1–385.

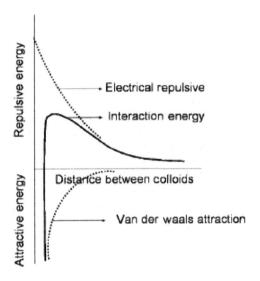

FIGURE 5.11

The net interaction energy is derived from the resulting combination of repulsive and attractive energy.

From Yu, H., Oduro, W., Tam, K., & Tsang, E. (2008). Chemical methods for preparation of nanoparticles in solution. Handbook of Metal Physics, 5, 1–385.

5.2.3 POLYMER NANOPARTICLES FOR SEMICONDUCTORS

Polymer matrices are embedded with metal in order to provide materials that possess the unique properties arising from nanoscopic size and shapes of metal clusters.

"The first approach for preparing the polymer-nanoparticle composites that are prepared by in-situ generation of metal nanoclusters by chemical reduction of a metal salt contained within a polymer matrix. The Polymer matrix would act as a template for their synthesis and would impart the necessary stability by providing a barrier against agglomeration of metallic nanoparticles which formed during and after the reduction process. The second approach towards polymer-nanoparticle composites involves the assembly of polymer scaffolds with performed monolayer protected clusters where, the surface monolayer prevents the metal clusters from agglomerating and provides good options for handling and processing" (Skaff et al., 2008).

Self-assembly of nanoparticles with polymers are providing access in order to stabilize metal and semiconductor nanocomposites for fabrication of new materials. However, many attributes of the individual building blocks like (Skaff et al., 2008):

- Size and shape of individual metal clusters
- Composition of the monolayer
- Functional groups on polymers would allow controlling the properties of these nanocomposites.

The control over the structure and morphology would be induced by changes in the polymer structure, where construction of composite materials by using assembly of nanoscopic building blocks (bottom–up) approach would provide a methodology compared to the top–down lithographic method. Moreover, the bottom–up approach would provide access to structures smaller and with higher 3D control than lithographic techniques such as EBL (Skaff et al., 2008).

5.2.3.1 Polymer nanoparticle blends

Mixing polymer materials and nanoparticles is the simplest method of composite formation, where it is accompanied by several aggregation problems. Moreover, the disparate nature of polymers and nanoparticles generally precludes homogenous particle dispersion such as the entropic gain associated with random dispersion is overwhelmed by enthalpy issues. Moreover, phase separation would destroy the nanoscopic integrity of nanoparticle, which would lead to a reduction in properties for the polymers. Many methods were considered to overcome and control the phase separation that occurs during the mixing processes (Skaff et al., 2008).

"Alivisatos and coworkers explored mixtures of CdSe nanoparticles and Poly-3-hexylthiophene (P3HT). A p-type conducting polymer for photovoltaic applications." P3HT and CdSe are dissolved in a mixture of chloroform and pyridine, with good solvents for the polymer and nanoparticle. Composite films would be obtained by spin casting these solutions to afford CdSe-based composites with particles contacts that are surmised to function as a percolating network for charge transport. Photovoltaic cells would generate from these materials an external quantum efficiency of 54% and a monochromatic power conversion efficiency of 6.9%. Also, efficiencies are lower than those of pure inorganic solar cells (Skaff et al., 2008).

> The polymer structure would be improved to better match the structure of conventional and aliphatic chain covered CdSe nanoparticles. Bawendi and coworkers blended poly(laurylmethacrylate) with (CdSe)ZnS core shell nanocrystals encapsulated with TOP. The aliphatic nature of the lauryl dodecylester side chains of this polymer serve to solubilize the nanoparticles by interaction is manifest in the optical clarity of the composites obtained upon blending, which suggests the likelihood of unaggregated nanocrystals in the composite material. Solid state quantum yields of these composites ranged from 22% to 44%.
>
> **Skaff et al., 2008.**

Another set of experiments were done by Schrock and coworkers. Functional polymers are prepared with pendant coordination sites capable of nanocrystal passivation.

> Monomeric norbornene derivatives are prepared with pendant phosphine oxides, where they are copolymerized with methltetracyclododecane through ring opening metathesis polymerization (ROMP) using molybdenum alkylidene catalyst to from a diblock copolymer with pendant coordination sites. This polymer is then dissolved in THF with CdSe nanocrystals to give optically clear solutions as the polymer coordinates to the nanocrystals. When this self assembly is performed in the presence of nanocrystals, the resulting TEM images show complete segregation of the nanocrystals into the coordinating block.
>
> **Skaff et al., 2008.**

5.2.3.2 Clay–nanoparticle polymer composites

The use of clay that comes from hydrated silicates of aluminum in combination with polymer materials would provide significant advantages in the physical properties compared to polymers alone. Other advances in composites in terms of melt processing would allow these materials to be prepared in the absence of any organic solvent (Skaff et al., 2008). Polypropylene–montmorillonite composites were prepared by varying the volume fraction of montmorillonite nanoparticles such as:

montmorillonite nanoparticles and varying block lengths of polypropylene-styrene or polypropylene-poly(methylmethacrylate) copolymers to probe mechanical and thermal properties. For example, a 3% inorganic loading into polypropylene gave a 30% increase in Young's modulus and a 30°C increase in heat deflection temperature when compared to native polypropylene.

Polymer-Nanoparticle Composites Part 1 (Nanotechnology), 2010.

Nonetheless, significant challenges are associated with blending polymers and nanoparticles in order to afford homogeneous, well-dispersed inorganic material within the polymer. In order for dispersion to be achieved, "the entropic penalty associated with addition of the nanoparticles must be balanced by favorable enthalpy interactions" (Polymer-Nanoparticle Composites Part 1 (Nanotechnology), 2010).

For this reason, polymer–clay hybrids composed of layered nanoparticles, such as silicates talc and mica which are aggregated to some degree, as the immiscibility of clay in the polymer leads to a very close proximity of sheets to one another. It is very important to mention that the degree of dispersion in these composites is generally referred to as:

- Unmixed highly aggregated
- Intercalated minimally aggregated
- Exfoliated well dispersed.

In intercalated cases, the polymer chains interpenetrate would be stacked silicate layers with small separation distances between layers. For the exfoliated or delaminated morphology, the silicate layers are well dispersed inside the polymer (Fig. 5.8). Exfoliation could be achieved by using the polar polymers, by the addition of a surfactant to the material, typically a long-chain alkylammonium salt. However, for nonpolar polymers such as poly(ethylene) and poly(propylene), the addition of a surfactant is not sufficient to overcome the entropic penalty. Thus a functional comonomer such as methyl methacrylate must be incorporated into the nonpolar polymer to allow nanoparticle dispersion within the matrix. Advances in processing have also led to decreased aggregation in clay–polyethylene materials, such as the use of supercritical CO_2 during polyethylene extrusion. Because the physical properties of these composites depend on the ability to produce controlled intercalation or exfoliation, the interest in small molecule and polymeric additives tuned to this target is growing significantly (Fig. 5.12).

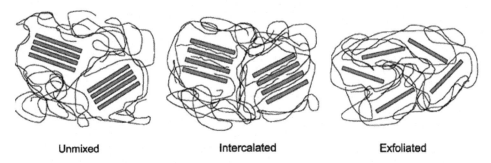

Unmixed   Intercalated   Exfoliated

FIGURE 5.12

Depiction of three types of clay–polymer hybrid materials showing different levels of particle dispersion.

From Polymer-Nanoparticle Composites Part 1 (Nanotechnology). (2010, May 25).
Retrieved 13.04.11, from What-When-How In depth information: http://what-when-how.com/nanoscience-and-nanotechnology/polymer-nanoparticle-composites-part-1-nanotechnology/.

5.3 APPLICATIONS

5.3.1 APPLICATION OF MAGNETIC NANOPARTICLES

Due to the advantages of size and shape, magnetic nanoparticles have a lot of applications in biomedicine, such as cell separation, drug carrier to cure cancer, and hyperthermia. The size of nanoparticles gives them the advantage of going into the body and can be divided into small scale to cure disease. The following section will focus on the cell separation and drug carrier function of nanoparticles.

5.3.1.1 Cell separation

In biomedicines, it is often necessary to separate blood and biological entities from an environment in order to analyze blood sample or body tissue. In magnetic nanoparticle, it can be used in these two processes: the identity of the biological entities from outside and the separation process due to different properties of entity and body tissue.

The following path demonstrates a process of cell separation (Fig. 5.13).

From Fig. 5.13, we can see that in part (A), the supernatant and the solution passing through the magnet at the beginning, while the supernatant is abort by the magnet and the desired solution can pass through without any supernatant in between. In part (B), it shows another way to abort supernatants, when it is in the external area, the magnetic field abort the supernatant to one side while the fluid can go through the area without supernatant involvement inside the solution, in this way, it can achieve a fluid flow fractionation. It can separate unwanted biomaterials from the magnetically tagged, which is gathering around the magnet due to the magnetic property.

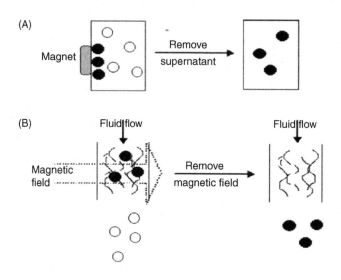

FIGURE 5.13

The standard methods of magnetic separation.

From Pankhurst, Q., Connolly, J., Jones, S., & Dobson, J. (2003). Application of magnetic nanoparticles in biomedicine. Journal of Physics D: Applied Physics.

With this kind of theory, it can be seen that it can be used in many parts of biomedical applications. It can separate rare tumor cells from blood and can be used to target the low-number cells. So it can lead to enhanced detection of materials in blood and processed technology for polymerase chain reactions. Also, for the separation applications, it can be used in optical sensing, due to the high sensitivity of the magnetic nanoparticles. If the target material is the solid matrix, it can act as the surface to locate the target material and to increase the concentration.

5.3.1.1.1 Drug delivery

Compared to traditional chemotherapies, magnetic nanoparticles have the advantages of treating the target area more specifically. Due to the guide of the magnet, it can be delivered to the specific site and then release the drug. Since the drug is used only for the specific site, so it can reduce the dose amount, so that it will reduce the effect of the systematic side effect on the body.

Many researches have been done on this subject. As early as late 1970s, experiments have been done on rat tails since rat tails have simplest structure and cam be analyzed more convenient. And then more experiments spread to swine, rabbits, and rats. But due to the limitation of technologies, not much of experiment has been done on human being, the only progress on human are the trail I phase experiment. With more and more successful experiments on animals, it is believed that technologies will be more mature to affect human beings.

Fig. 5.14 shows the mechanism of drug delivery application of magnetic nanoparticles.

From Fig. 5.14, it can be seen that the drug is first coated with magnetic material and then injected into the body. The external magnet guides the nanoparticles into one specific site along the blood vessel. When it arrives at the target site, it can be attracted into the tumor tissue. It then releases the drug due to the change of the internal environment (such as the pH). In previous studies, much research has been done on rabbits (Pankhurst, Connolly, Jones, & Dobson, 2003). It shows no toxicity of the drug delivery group since the drug was delivered into the tumor directly for treatment. The whole process lasted 30 days and the result is encouraging and shows that magnetic nanoparticles can be used in this kind of drug delivery.

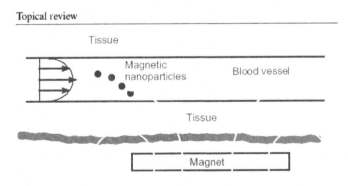

FIGURE 5.14

Hypothetical magnetic drug delivery system in cross section.

From Pankhurst, Q., Connolly, J., Jones, S., & Dobson, J. (2003). Application of magnetic nanoparticles in biomedicine. Journal of Physics D: Applied Physics.

5.3.2 APPLICATION OF METALLIC NANOPARTICLES

Metallic nanoparticles have different applications in different types of fields. These particles have structure with optical and electronic properties. These particles can be used in electronic applications, as an example the positioning of individual molecules of lambda-DNA in an electrode gap (Sarma, 2003).

Metal nanoparticles have recently found to be applicable in catalysis. Haruta (2004) have discovered that supported gold nanoparticles are extremely catalytically active for carbon monoxide (CO) at the temperature much lower than room temperature, when the particle size of the gold is less than 5 nm. This finding is very efficient for nanocatalysts for many chemical reactions. The interfaces created between the gold nanoparticles with transition metal oxide supports, such as zirconia, iron oxide, titania, cerium oxide, and others, are attributed to account for the ultra-high catalytic activity and selectivity. However, the activity of these metal clusters (in particular, gold nanoparticles) in a reaction is very much dependent on the method of preparation, particle size, shape, dispersion, and the type of supported metal oxide (Yu et al., 2008).

One of the most applicable metal nanoparticles is gold nanoparticles which have been used in different fields. These applications are enzymatic biosensor, genosensors, immunosensors, and electrocatalytic chemosensors (Fredy, 2008).

5.3.3 APPLICATION OF POLYMER NANOPARTICLES

Unique properties for polymers that include their thermal behavior, processability, and ability to assemble into ordered structures would offer potential for compatibilizing nanocrystals, directing their assembly, and providing a mechanism for charge transports. The confirmation of theoretical predictions for spatial distribution for nanoparticles inside polymer hosts would depend on interactions between ligand periphery and polymer environment (Skaff et al., 2008).

A nonexhaustive review of some recent advances for semiconductors; where four major approaches have designed to integrate semiconductor nanoparticles into polymer matrices (Skaff et al., 2008):

- Simple mixing of nanoparticles with functional or nonfunctional polymers
- Growth of nanoparticles from organometallic precursors
- Chain-end attachment of polymers to nanoparticle surfaces, either by mixing end functional polymers with the nanoparticles or by growing polymers radially outward from nanoparticle surface
- Assembling nanoparticles in polymer templates, where the structure of template dictates the assembly of nanoparticles.

The integration of nanoparticles into polymers is one of the most significant theoretical and experimental interests in the polymer and engineering communities.

Inorganic fillers have been used for some time in conjunction with organic polymer materials, largely in an effort to enhance the physical and mechanical properties over those of the polymers alone.

Polymer-Nanoparticle Composites Part 1 (Nanotechnology), 2010.

In the 19th century, huge research efforts were done by Charles and Nelson Goodyear, the pioneers in the chemistry of rubber, who had showed that vulcanized rubber can be toughened significantly by the addition of Zinc Oxide (ZnO) and Magnesium Sulfate (MgSO$_4$). Also, Leo Baekeland investigated the use of Silicate clay in phenol resins that helped Bakelite, where the first mass produced synthetic polymer composite, become a commercial success.

Many additional commodity materials were developed based on the enhanced properties that result from filling rubber with clay. Recently, it was found that rubber particles embedded in Nylon and other polymers afford composites with outstanding impact resistances. The research and scientific efforts have been focusing on the integration of varieties of nanoparticles into polymers and to study the impact of these composites over a broad spectrum of applications, from Polymer Engineering to advanced Electronic Materials (Polymer-Nanoparticle Composites Part 1 (Nanotechnology), 2010).

Tremendous advances in characterization technology is a common thread through all disciplines in nanotechnology, where nanoscopic objects and materials can be visualized with greatly improved resolution relative to only in the last 20 years ago.

> Fig. 2 illustrates a few examples of nanomaterials with dimensions of 1 micron and less. In the nanoparticle-polymer area, current efforts now reach far beyond (or below) the use of micron-sized particle fillers (e.g., layered silicates) and into very small metallic and semiconductor nanoparticles as small as 1–2 nm in diameter.
>
> **Polymer-Nanoparticle Composites Part 1 (Nanotechnology), 2010.**

A discussion of clay-based nanocomposites will be followed by a focus on metallic and semiconductor-based hybrid materials. The use of polymers as a means to provide exquisite order to nanoparticles will be described based on the ability of polymer materials to assemble into nanostructures. Finally, potential applications of polymer–nanoparticle composites would be discussed, with a special focus on the use of dendrite polymers and nanoparticles for catalysis (Polymer-Nanoparticle Composites Part 1 (Nanotechnology), 2010) (Fig. 5.15).

The incorporation of these nanoparticles into polymer matrices has shown great potential for photovoltaic cells, LEDs, tunable lasers, and biologically active tags and diagnostic devices.

5.4 CURRENT RESEARCH AND FUTURE PERSPECTIVES

Nanoparticles are very important in the field of nanotechnology, and thus there is much research about their synthesis and applications. Magnetic particles are an important subset with many potential uses. Biomedicine is a particular field where nanoparticles are of much importance. Magnetic nanoparticles (MNPs) have been increasingly researched for their uses in this area. Pankhurst et al. (2003) reviewed some of the physical properties of magnetic nanoparticles and described some of their biomedical applications. The general mechanism of magnetic nanoparticles for drug delivery is demonstrated in Fig. 5.16. We can see that an external magnet is the key to the moving of magnetic nanoparticles along the blood vessel. When they arrive at the targeted site, the environment of the site breaks the coating of the nanoparticles and releases the drug.

There have been many publications devoted to this magnetic nanoparticle drug delivery system. Alexiou et al. (2000) studied the injection of magnetic particles (ferrofluids) to bind anticancer agents to a tumor area with magnetic field. The model used in their experiments is rabbits with squamous cell carcinoma. Alexiou found that ferrofluid mitoxantrone had great effect on transporting anticancer drug to the tumor and had the effect of permanent remission of the squamous cell carcinoma compared to a no-treatment group. They used starch to coat the drug such that it binds to amine groups of MTX-HCl at a pH of 7.4.

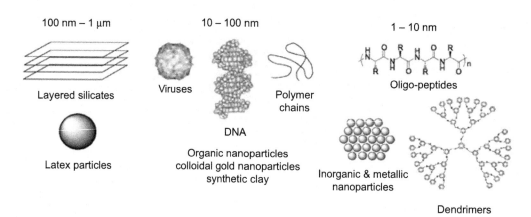

100 nm – 1 μm

Layered silicates

Latex particles

10 – 100 nm

Viruses

DNA

Polymer chains

Organic nanoparticles colloidal gold nanoparticles synthetic clay

Inorganic & metallic nanoparticles

1 – 10 nm

Oligo-peptides

Dendrimers

FIGURE 5.15

Building blocks for nanotechnology.

From Polymer-Nanoparticle Composites Part 1 (Nanotechnology). (2010, May 25). Retrieved 13.04.11, from What-When-How In depth information: http://what-when-how.com/nanoscience-and-nanotechnology/polymer-nanoparticle-composites-part-1-nanotechnology/.

Many studies are dedicated to modifying these particles. Park, Kim, Lim, and Kim (2008) have tested the change in size and saturation magnetization of magnetic nanoparticles with different lecithin concentrations. For this purpose, they first made lecithin-adsorbed superparamagnetic nanoparticles through three steps. First, iron nanoparticles were formed by thermal decomposition of $Fe(CO)_5$ compound. In order to obtain magnetite particles, their mild oxidation was conducted by trimethylamine N-oxide. During ultrasonication, toluene, ethyl alcohol, and distilled water were used to wash the obtained solution. The washed black powders were added in 80 mL phosphate buffered saline (PBS) solution at 80°C to adsorb lecithin. During that, ultrasonic holding time is changed from 0.5 to 1.5 h and the reacted lecithin concentration for dispersing particles was also changed from 0% to 50% (w/v). The colloids were centrifuged at 700 rpm for 5 min after they cooled to room temperature. Then, particles were dried in a vacuum and characterized by different methods. It was observed that there is change in particle size from 0% to 50% concentration of lecithin. The average size of pure magnetic particles without lecithin adsorption was 9.82 nm, and the average particle size increased up to 12.81 nm in lecithin concentration of 20% (w/v). However, the distinguishable change in particle size was not observed in the concentration of above 20% (w/v). Furthermore, as lecithin concentration increased from 0% to 50% (w/v), the saturation magnetization decreased from 76 to 54 emu/g nonlinearly (Fig. 5.16A). The decrease in saturation can be described by the thickening of the shell layer. The increase of nonmagnetic volume fraction, which was demonstrated by the weight loss measurement during heating of magnetic particles shown in Fig. 5.16B, which illustrates the percent weight loss with the increasing temperature for the adsorbed magnetic nanoparticles with different lecithin concentrations.

The cell viability was also represented in in vitro test results when the concentration of magnetic colloid changed. From this result, the magnetic colloids with the concentration of 32 μg/mL that showed the relatively superior cell viability and had the high volume compared to cell numbers were chosen as fluid samples for in vivo tests.

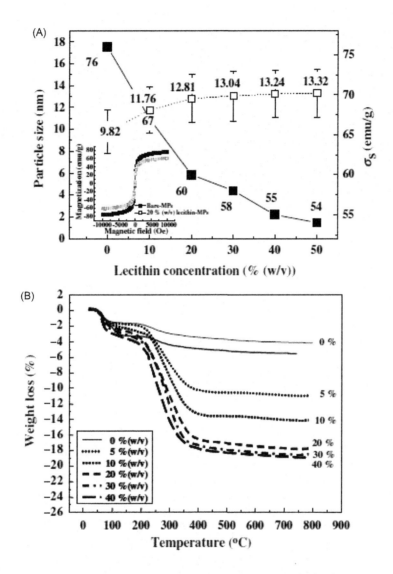

FIGURE 5.16

(A) Lecithin concentration versus particle size and (B) temperature versus percentage of weight loss.

From Park, S., Kim, H., Lim, H., & Kim, C. O. (2008). Surface-modified magnetic nanoparticles with lecithin for applications in biomedicine. Current Applied Physics, 8, 706–709.

In addition to looking at the applications of magnetic nanoparticles, much research is dedicated to the actual preparation of the particles. Roca, Costo, Rebolledo, and Veintemillas (2009) have proposed the method of synthetic routes to produce ferrites with different sizes and shapes, which are the most suitable materials for biomedical applications. The applications of magnetic nanoparticles in medicine

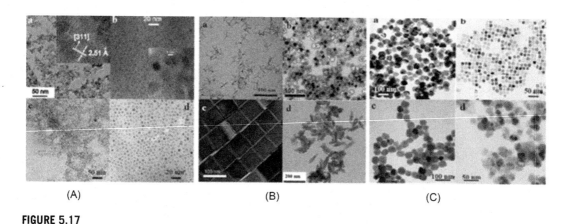

(A)   (B)   (C)

FIGURE 5.17

(A) Small particle sizes with four different methods. (B) Medium particle sizes with different shapes. (C) Large particle sizes obtained with different methods.

From Roca, A., Costo, R., Rebolledo, F., & Veintemillas, S. (2009). Progress in the preparation of magnetic nanoparticles for applications in biomedicine. Journal of Physics D: Applied Physics, 42, 11.

mostly depend on their size and compositions. They have divided the magnetic nanoparticles with sizes below 10 nm, sizes between 10 and 30 nm, and sizes around the monodomain–multidomain magnetic transition (>30 nm). They presented different methods for each particle size. For synthesizing small particles (<10 nm) or ultrasmall superparamagnetic iron oxide particles, four different methods have been provided: the Massart method, arrested precipitation method, precipitation in reverse microemulsions, and thermal decomposition of organometallic compounds. High-temperature decomposition of organic precursors and the higher concentration of the iron precursor, higher Fe/oleic acid ratio, and the use of high-boiling-point solvents such as octyl ether or trioctylamine lead to particles with sizes over 10 nm (called superparamagnetic particles) or medium size particles (10–30 nm). The Fe/surfactant ratio also affects the final nanoparticle size. Large particles (>30 nm) can be synthesized by controlling the ratio of ferrous and ferric salts with hexanediamine. The classical coprecipitation method can produce particles from 9 to 40 nm. Magnetite nanoparticles with sizes between 20 and 30 nm have been obtained by Fe electro-oxidation in the presence of an amine surfactant, which acts both as a supporting electrolyte and coating agent. They can also be synthesized by thermal decomposition of organic precursors in organic media. Magnetite particles up to 50 nm in diameter can be obtained by decreasing the solvent coordination capacity (Fig. 5.17).

Magnetic nanoparticles have been combined with other materials resulting in systems with multifunctional capabilities for sensing, biocatalysis, targeted infection, magnetic resonance imaging, drug delivery, etc. They also described examples of the preparation of multicomponent systems with purely inorganic or organic–inorganic characteristics.

A method of synthesis, the hydrothermal method, has recently emerged as an effective method for the synthesis of zinc oxide nanostructures. Zinc oxide is a semiconductor used for a variety of purposes, such as sensors. ZnO nanoparticles and nanorods were successfully synthesized by a hydrothermal method. Cetyltrimethyl ammonium bromide (CTAB) surfactant plays an important role in morphological

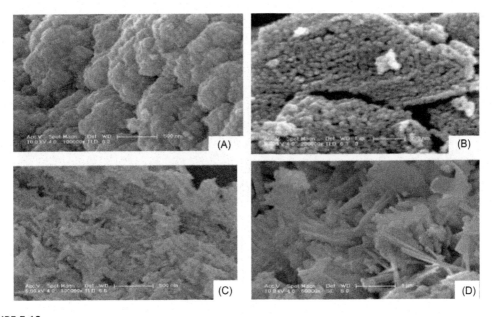

FIGURE 5.18

SEM images of ZnO nanoparticles (A) the absence of CTAB, (B) pH = 3, (C) pH = 7, (D) pH = 9.

From Sridevi, D., & Rajendran, K. (2009). Preparation of ZnO nanoparticles and nanorods by using CTAB assisted hydrothermal method. International Journal of Nanotechnology and Applications, 43–48.

changes during the hydrothermal method. The nanostructures have been characterized by X-ray diffraction (XRD), energy dispersive X-ray analysis, and scanning electron microscopy (SEM). Moreover, the effect of pH on morphology and photoluminescence properties is discussed (Fig. 5.18).

To prepare the experiment: 0.02M of $ZnCl_2 \cdot 2H_2O$ and 0.02M of CTAB were added to 50 mL of water and an appropriate amount of NaOH was added in order to have the pH value of the solution to be 3, 7, and 9. The solutions would be stirred for half an hour and poured into a stainless Teflon-lined 100 mL autoclave to maintain at 1300°C for 24 h. Later on, the mixture would be cooled to room temperature, the resulting precipitate would be collected and washed with ethanol and annealed at 4000°C for 1 h. This experiment successfully prepared ZnO nanoparticles and nanorods by a CTAB-assisted hydrothermal approach. Experiments showed that precursors and pH would influence morphologies and photoluminescence properties (Sridevi & Rajendran, 2009).

There are other methods for synthesizing zinc oxide (ZnO). For example, zinc oxide has been formed by the reaction of zinc metal with methanol at very low temperatures. This experiment was performed in order to synthesize zinc oxide nanoparticles through a simple and soft reaction of the zinc metal with methanol (CH_3OH) at 300°C. However, the nanoparticles' diameters would range from 50 to 200 nm with an average diameter of 100 nm. Also, this reaction would involve the cleavage of C–O bond of the methanol, which occurs readily on the Zn surface. It is important to mention that addition of the ethylenediamine to this reaction yields nanorods implying that it acted as a shape-controlling agent.

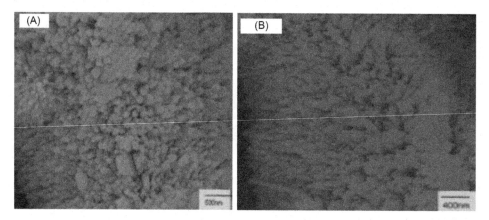

FIGURE 5.19

The FESEM images of nanoparticles of ZnO prepared by the reaction of zinc metal with methanol at 300°C, respectively.

From Hua, G., Zhang, L., Dai, J., Hu, L., Dai, S. (2011). Synthesis of dimension-tunable ZnO nanostructures via the design of zinc hydroxide precursors, Applied Physics A, 102(2), 275–280, with permission of Springer.

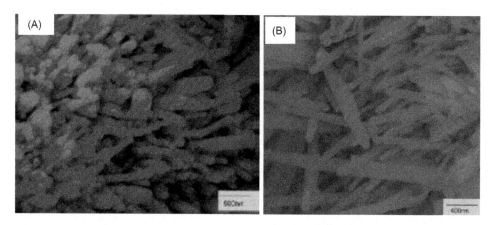

FIGURE 5.20

The FESEM images of nanorods of ZnO prepared by the reaction of zinc metal with methanol by the addition of ethylenediamine at 300°C.

From Hua, G., Zhang, L., Dai, J., Hu, L., Dai, S. (2011). Synthesis of dimension-tunable ZnO nanostructures via the design of zinc hydroxide precursors, Applied Physics A, 102(2), 275–280, with permission of Springer.

"A plausible mechanism is proposed for the formation of these nanoparticles and it is expected that this synthetic technique can be extended to obtain other metal oxides" (Shah, 2008) (Figs. 5.19 and 5.20).

The reaction of the methanol (solvent) with Zn powder provides a simple "route for the preparation of ZnO nanoparticles." This route may provide an effective low-cost approach that should promise a future large-scale synthesis of ZnO nanostructures for applications in nanotechnology (Shah, 2008).

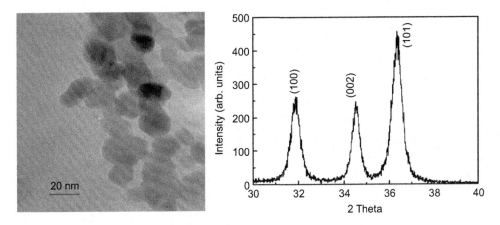

FIGURE 5.21

TEM image and XRD pattern of ZnO nanoparticles.

From McCluskey, W. M. (2005). Infrared spectroscopy of impurities in ZnO nanoparticles. Materials Research Society.

Because ZnO nanostructures have many uses, it is important to understand their exact composition. The impurity in ZnO nanoparticles was investigated with infrared (IR) spectroscopy, and the results show the presence of CO_2 impurities with the ZnO nanoparticles. For this reason, an isotopic substitution had been used in order to verify "the frequency assignment and the results demonstrate conclusively that the impurities originate from the precursors." Also, an isochronal annealing experiment was performed in order to study the formation and stability of the CO_2 molecules. From this article, ZnO nanoparticles were synthesized by reaction of zinc acetatedehydrate, $Zn(CH_3COO)_2\cdot2H_2O$, with sodium hydrogen carbonate, $NaHCO_3$. The accuracy for this experiment is very important; for this reason and in order to prevent any contamination that would cause by the ambient; this mixture would be sealed in an Ar-gas-filled quartz ampoule. The TEM image and XRD pattern of the nanoparticles are shown in Fig. 5.21. The sizes of the particles are about ~15 nm and XRD pattern confirms the wurtzite structure. The powder was pressed into thin pellets (~0.25 mm) to perform the IR spectroscopy (McCluskey, 2005).

Results from the isochronal annealing show that CO_2 molecules are formed by thermal decomposition of reaction products and are believed to be trapped within the particles.

When using such particles, it is important to be able to understand how to control their sizes and the effect that different sizes have. Nishio, Ikeda, Gokon, Tsubouchi, and Na (2007) prepared MNPs with sharply dispersed sizes ranging between 30 and 100 nm in aqueous solution at relatively low temperatures (4–37°C) without any surfactant. For this purpose, sodium nitrate ($NaNO_3$) (0.75 g, 8.80 mmol) was added as an oxidant into a 21 mM aqueous NaOH solution (475 mL, pH 12-13). Deaerated 0.1M ferrous chloride aqueous solution (25 mL) was added to the alkaline solution, which was kept at 4, 15, 25, and 37°C for 24 h. MNPs thus synthesized were separated from the solution with a magnet and washed several times with pure water. The morphology/size, crystal phase, and magnetization for the MNPs using TEM, X-ray diffractometer, and VSM, respectively, were investigated.

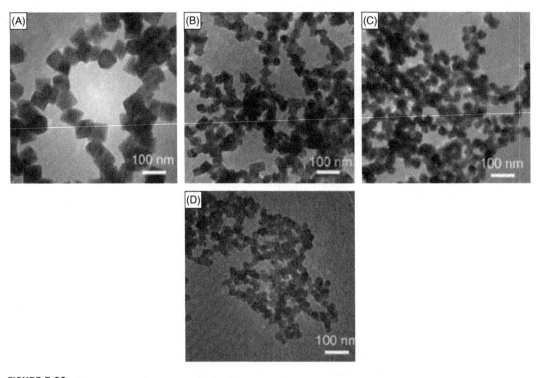

FIGURE 5.22

TEM images for MNPs synthesized at (A) 4°C, (B) 15°C, (C) 25°C, and (D) 37°C.

From Nishio, K., Ikeda, M., Gokon, N., Tsubouchi, S., & Na, H. (2007). Preparation of size-controlled (30–100 nm) magnetite nanoparticles for biomedical applications. Journal of Magnetism and Magnetic Materials, 310, 2408–2410.

The nanoparticles were essentially magnetite because they had a lattice constant quite near to that reported for bulk Fe_3O_4. All the MNPs had a residual magnetization because their sizes were larger than the superparamagnetic limit (26 nm) of Fe_3O4.

Fig. 5.22 shows that increasing the synthesis temperature from 4°C to 37°C leads to a decrease in the MNP's size (from 10275.6 to 31.7 nm) and changes the MNP's shape (from octahedral to nearly spherical). Furthermore, the size distribution of our synthesized MNPs becomes wide. The size distribution increases from 5.5% to 15.5%. This indicates that a higher reaction temperature accelerates both nucleation and growth of the magnetite crystals in the aqueous solution.

Goodwin and his colleagues studied the single-dose toxicity effect on the treatment of doxorubicin coupled to novel magnetically targeted drug carrier. Using swine as model, they divided 18 swine into 6 groups to study the differences between without magnetic field and with magnetic field. MTC-DOX (magnetically targeted drug carrier—doxorubicin) was injected into models to study the site-specific delivery function. By differentiating dose concentration and existence of magnetic field, they discovered that high concentration of dose without magnetic field would have serious clinical signs, otherwise no clinical signs were observed. In the side of pathology, the MTC particles were clearly shown in a photomicrograph (Fig. 5.23).

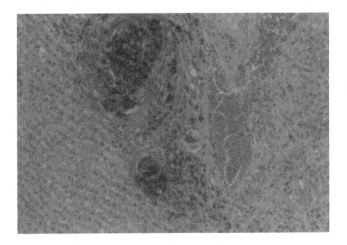

FIGURE 5.23

Photomicrograph illustrating portal area of liver from animal in the MTC control group.

From Scott C. Goodwin, Craig A. Bittner, Caryn L. Peterson, Gordon Wong. Single Dose Toxicity Study of Hepatic Intra-arterial Infusion of Doxorubicin Coupled to a Novel Magnetically Targeted Drug Carrier. Toxicol. Sci. (2001) 60 (1): 177–183.

Fig. 5.23 shows clearly that the MTC particles (black cluster) are accumulating around the tumor (red part (light gray in print version)). With the help of magnetic field, they managed to release the nanoparticles around the tumor and none in circulatory system, according to toxic kinetic analysis.

5.5 CONCLUSIONS

Nanocomposite materials fabricated using polymer-mediated self-assembly of nanoparticles is a versatile and effective tool for the advancement of nanotechnology. The ability to control properties of the polymer and the nanoparticle core and shell at the molecular level provides access to a wide variety of materials with tunable properties. Nanoparticles in polymers would provide a powerful platform for device fabrication.

NANOPOLYMERS[*]

6.1 NANOPOLYMERS AND THEIR APPLICATIONS

Polymers are classified as amorphous, crystalline, or semicrystalline. Amorphous polymers are non-ordered and usually transparent, like the polymers used in plastic bottles to hold water. Nonordered (amorphous) polymers have lower glass transition temperatures than semicrystalline polymers and crystalline polymers. Thus, it is easier to process amorphous polymers. Semicrystalline polymers have some ordered regions and some nonordered regions that correspondingly represent crystalline and amorphous regions in polymers' matrices. Semicrystalline polymers possess a higher glass transition temperature than amorphous polymers and are stronger than amorphous polymers, due to the arrangement of the molecules. Crystalline polymers are highly ordered polymers with significantly stronger mechanical properties, compared to semicrystalline polymers. This is because crystalline polymers form a crystal lattice that enables the polymer to resist different types of forces.

When polymers are fabricated as nanopolymers, they exhibit different properties. This is because nanopolymers are significantly smaller than macromolecular polymers. This allows nanopolymers to physically interact with their surroundings in a different manner. For example, consider two polymers: a centimeter-sized polymer and a nanopolymer. If a shear force is put on each polymer respectively, the nanopolymer would be able to resist the shear force. This can be imagined by considering a lever fulcrum system.

The potential applications of nanopolymers are wide diverse. Nanopolymers can literally replace every application of polymers. This includes various markets such as defense, telecommunications, basic materials (assisting the process), household goods, a vast majority of services (used in daily operation), and utilities. This can be broken down further to include toothpaste, plastic containers, etc.

Nanopolymers are used because they have very desirable chemical properties, namely: high tensile strength, good chemical resistivity, as well as being able to contain metals and other compounds. For example, the nanopolymer could contain highly conducting materials to develop a nano-sized circuit. In addition, nanopolymers can be fabricated from various structures. Some self-assembled structures include lamellar, lamellar-within-spherical, lamellar-within-cylindrical, cylindrical-within-lamellar, and the spherical-within-lamellar geometry. For non-self-assembled structures, this includes dendrimers, hyperbranched polymers, polymer brushes, nanofibers, polymeric nanotubes, and polymeric nanocapsules (Biomaterial Laboratory of the Universidad Politecnica de Madrid). The above mentioned nanostructures will only be listed, as a thorough explanation with diagrams is beyond the scope of the chapter.

[*] By Yaser Dahman, Kevin Deonanan, Timothy Dontsos, and Andrew Iammatteo.

Nanotechnology and Functional Materials for Engineers. DOI: http://dx.doi.org/10.1016/B978-0-323-51256-5.00006-X

6.2 FABRICATING NANOPOLYMERS: ELECTROSPINNING

The electrospinning process is commonly used as a modern nanofiber fabrication technique. Other fabrication techniques include drawing, template synthesis, phase separation, and self-assembly. A schematic of the process is shown in Fig. 6.1.

The ability to generate polymer fibers using an electrically charged jet was discovered in the 1930s. Since this time, the electrospinning process has developed and can now be manipulated to a large degree.

The process proceeds by having a polymer solution with the desired nanofiber polymer mixed with a solvent in a syringe being pumped through a nozzle. A high direct current (DC) electric potential is applied between a collecting plate and the syringe. The solution moves toward the collector. A Taylor cone is formed at the meniscus of the solution, from which the solution proceeds toward the grounded target. After the solution is elongated, whipping occurs where the solution starts to move in random circular directions. The solvent evaporates during the whipping process and the fiber is deposited as a solid polymer. The resulting polymer is a highly porous non-woven mat (Haghi, 2009).

There are many parameters that can be manipulated in order to get the desired thickness and composition including the properties of the polymer itself (molecular weight and architecture), solution properties, electric potential, distance to the plate, ambient parameters, and the motion of the target.

6.3 ENVIRONMENTAL APPLICATIONS

Nanotechnology is still an up-and-coming field, with many of the applications still being researched. Nanofibers, mainly fabricated by electrospinning, have exhibited great potential for many emerging environmental applications. They can be considered as one of the safest nanomaterials due to their extremely long length and their ability to be embedded within other media. Their high surface-to-volume ratio, large porosity (up to over 80%), and adjustable functionality are also much more

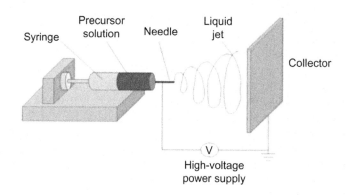

FIGURE 6.1

Electrospinning setup schematic.

From Nanotechweb.org, 2011.

effective than conventional non-woven and polymeric membranes in particulate separation and liquid filtration. Technology advances such as multiple-jet electrospinning and electroblowing for mass production of nanofibers have made it practical to use nanofibrous scaffolds as a unique and breakthrough component in separation media for both gas and liquid filtration.

The majority of current electrospinning related studies are concerned with the generation of new nano-structured materials and their applications, encompassing biological membranes (substrates for tissue regeneration, immobilized enzymes and catalyst systems, wound dressing articles, artificial blood vessels and material for the prevention of postoperative induced adhesions), aerosol filters and clothing membranes for protection against environmental elements, and optical and chemical sensors and electrical conductors. The applications discussed in this section will be focused on membranes for fuel cells, heavy ion removal in industrial wastewater and solar energy.

6.3.1 FUEL CELLS

6.3.1.1 Introduction

As their name suggests, fuel cells convert chemical energy from a fuel source into electrical energy. More specifically, the electricity is generated from the oxidation reaction between the fuel supply and an oxidizing agent. Thus, the amount of energy produced by a fuel cell is limited to the supply of reactant and oxidant.

Fuel cells represent a *thermodynamically open* system, such that they consume a reactant from an external source which must be replenished. Conventional batteries store electric energy chemically and represent a *thermodynamically closed* system.

There are many different combinations of fuels and oxidants. A common one is the hydrogen fuel cell, which utilizes hydrogen as its fuel and air as its oxidant. Fig. 6.2 shows a model of a direct methanol fuel cell (DMFC), where the actual fuel cell stack is the layered cube in the center of the image.

6.3.1.2 Theoretical background

Although there are a variety of fuel cells, they all work in the same general manner. Basically, there are three main components to any fuel cell: the anode, the cathode, and the electrolyte. This is shown in Fig. 6.3.

Two chemical reactions take place in this system, resulting in the consumption of fuel and production of water or carbon dioxide. This generates an electrical current. At the anode, the fuel is oxidized via interaction with a catalyst turning it into a positively charged ion. The ions then pass through the electrolyte toward the cathode, leaving their electrons behind. This occurs because the electrolyte is designed to block electrons so that they may travel to the cathode through a wire and create an electric current. In the cathode, the ions are reunited with the electrons and react with another chemical to create water or carbon dioxide (Larminie, 2003).

In a proton-exchange membrane fuel cell, a solid, porous membrane (the electrolyte) separates the anode and the cathode and must simultaneously facilitate rapid proton transport as well as act as an effective fuel barrier. Regarding DMFCs, it is of great interest to find novel and high-performance polymer electrolyte membranes with high fuel barrier characteristics.

Shabani, Hasani-Sadrabadi, Haddadi-Asl, and Soleimani (2010) explored the applications of polymeric nanofibers in fuel cells. More specifically, poly(ether sulfone) (PES) underwent sulfonation to produce sulfonated poly(ether sulfone) (SPES). Nanofibrous membranes were then constructed via

FIGURE 6.2

Working model of a DMFC.

From Wikipedia.

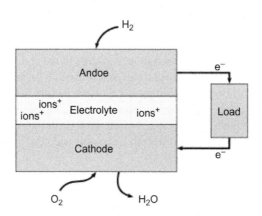

FIGURE 6.3

Basic configuration of a fuel cell.

From Wikipedia.

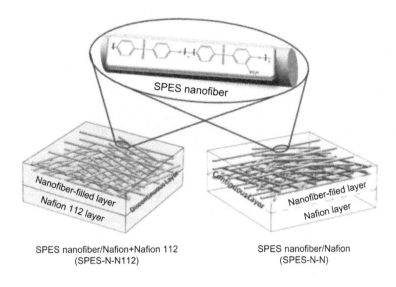

SPES nanofiber

Nanofiber-filled layer
Nafion 112 layer

Nanofiber-filed layer
Nafion layer

SPES nanofiber/Nafion+Nafion 112
(SPES-N-N112)

SPES nanofiber/Nafion
(SPES-N-N)

FIGURE 6.4

Designated bilayer membranes.

From Shabani, I., Hasani-Sadrabadi, M. M., Haddadi-Asl, V., & Soleimani, M. (2010). Nanofiber-based polyelectrolytes as novel membranes for fuel cell applications. Journal of Membrane Science, 233–240.

electrospinning. The resulting porous membranes were then impregnated with a solution of Nafion, allowing the pores to fill and a uniform top layer to form. Another type of SPES membrane was formed by electrospinning on a Nafion112 membrane surface and impregnating it with Nafion as previously described. This produced both SPES-N-N and SPES-N-N112. After undergoing annealing at 12°C for 12 h and several chemical baths, the membranes were tested for proton conductivity and methanol permeability. Fig. 6.4 shows a representation of the produced polynanofiber membranes.

The membranes have a higher selectivity due to the incorporation of SPES nanofibers into the Nafion matrix as well as increased open-circuit voltages. Each of these factors allows for these membranes to be utilized as promising polyelectrolyte membranes for fuel cells.

6.3.2 HEAVY METAL ION REMOVAL

6.3.2.1 Introduction

The removal of toxic metals from wastewater is of great concern due to its high impact on the environment and public health. This problem can be solved by utilizing adsorption technology, such as ion-exchange resins, activated charcoal, and ion-chelating agents. This has been an improvement over conventional precipitation methods which can re-pollute the water due to difficulties in recovery.

The removal and recovery of heavy metals from wastewater has been categorized by various chemical processes. The four major classes of chemical separation technologies are chemical precipitation, electrolytic recovery, adsorption/ion exchange, and solvent extraction/liquid membrane separation. Fig. 6.5 illustrates these chemical separation techniques and categorizes them based on specific types of treatment.

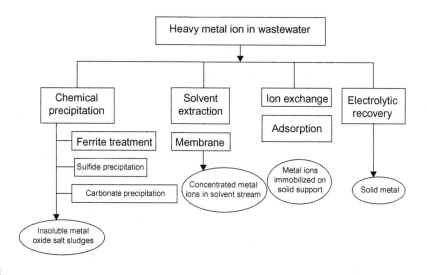

FIGURE 6.5

Various chemical treatment methods for heavy metal removal from wastewater.

From Lewinsky, A. A. (2007). Hazardous materials and wastewater: Treatment, removal and analysis. New York: Nova Science Publishers Inc.

6.3.2.2 *Processes of removal*

6.3.2.2.1 Chemical precipitation

Chemical precipitation is the most common technology used in removing dissolved (ionic) metals from solutions, such as process wastewaters containing toxic metals. The ionic metals are converted to an insoluble form (particle) by the chemical reaction between the soluble metal compounds and the precipitating reagent. The particles formed by this reaction are removed from solution by settling and/or filtration. The unit operations typically required in this technology include neutralization, precipitation, coagulation/flocculation, solids/liquid separation, and dewatering.

The effectiveness of a chemical precipitation process is dependent on several factors, including the type and concentration of ionic metals present in solution, the precipitant used, the reaction conditions (especially the pH of the solution), and the presence of other constituents that may inhibit the precipitation reaction.

The most widely used chemical precipitation process is hydroxide precipitation (also referred to as precipitation by pH), in which metal hydroxides are formed by using calcium hydroxide (lime) or sodium hydroxide (caustic) as the precipitant. Each dissolved metal has a distinct pH value at which the optimum hydroxide precipitation occurs—from 7.5 for chromium to 11.0 for cadmium. Metal hydroxides are amphoteric, which means that they are increasingly soluble at both low and high pH values. Therefore, the optimum pH for precipitation of one metal may cause another metal to solubilize or start to go back into solution. Most process wastewaters contain mixed metals and so precipitating these different metals as hydroxides can be a tricky process (Lewinsky, 2007).

6.3.2.2.2 Solvent extraction

Solvent extraction is a common form of chemical extraction using organic solvent. It is commonly used in combination with other technologies, such as solidification/stabilization, incineration, or soil washing, depending upon site-specific conditions. Solvent extraction also can be used as a stand-alone technology in some instances. Organically bound metals can be extracted along with the target organic contaminants, thereby creating residuals with special handling requirements. Traces of solvent may remain within the treated soil matrix, so the toxicity of the solvent is an important consideration. The treated media are usually returned to the site after having met best demonstrated available technology and other standards.

It has also been shown to be effective in treating sediments, sludge, and soils containing primarily organic contaminants such as PCBs, VOCs, halogenated solvents, and petroleum wastes. The process has been shown to be applicable for the separation of the organic contaminants in paint wastes, synthetic rubber process wastes, coal tar wastes, drilling muds, wood-treating wastes, separation sludge, pesticide/insecticide wastes, and petroleum refinery oily wastes (Lewinsky, 2007).

6.3.2.2.3 Ion exchange

Ion exchange is a reversible chemical reaction wherein an ion (an atom or a molecule that has lost or gained an electron and thus acquired an electrical charge) from a wastewater solution is exchanged for a similarly charged ion attached to an immobile solid particle. These solid ion-exchange particles are either naturally occurring inorganic zeolites or synthetically produced organic resins. The synthetic organic resins are the predominant type used today because their characteristics can be tailored to specific applications (Lewinsky, 2007).

An organic ion-exchange resin is composed of high-molecular-weight polyelectrolytes that can exchange their mobile ions for ions of a similar charge from the wastewater. Each resin has a distinct number of mobile ion sites that set the maximum quantity of exchanges per unit of resin. Table 6.1 summarizes some of the various components of the ion-exchange process.

6.3.2.2.4 Electrolytic recovery

This process utilizes what is called the electrolytic cell to recover heavy metal ions from wastewater. The cell is composed of an anode and a cathode submerged in an electrolyte. When a current is applied, dissolved metals in the electrolyte are reduced and deposited on the cathode. One advantage of this process is that it can target specific contaminants in the wastewater without the addition of chemicals that can produce a large amount of sludge. Through this process the metal is often reusable, defining this as a "recovery" process as opposed to an end-of-pipe process.

Electrolytic recovery is not a useful method for all contaminants. It is most effective in removing the noble metals, such as gold and silver because of their high electrode potential and ease of being reduced and deposited onto the cathode. Metals such as aluminum and magnesium which favor oxidation and have lower electrode potentials cannot be removed by this process. For metals such as copper and tin, a higher current must be applied for this method to be utilized. The following chart is an illustration of the electrolytic recovery process (Lewinsky, 2007) (Fig. 6.6).

The application of nanofibers to the removal of heavy ions from wastewater was explored by Teng, Wang, Li, and Zhang (2010) using a thioether-functionalized organic–inorganic composite membrane with mesostructure. The aforementioned thioether was developed using a combination of a sol–gel process and electrospinning. The film that resulted was fabricated using polyvinylpyrrol-iodone

Table 6.1 Selectivity of Ion-Exchange Resins in Order of Decreasing Preference

Strong Acid Cation Exchanger	Strong Base Anion Exchanger
Barium	Iodide
Lead	Nitrate
Calcium	Bisulfite
Nickel	Chloride
Cadmium	Cyanide
Copper	Bicarbonate
Zinc	Hydroxide
Magnesium	Fluoride
Potassium	Sulfate
Ammonia	
Sodium	
Hydrogen	

Reproduced from Table 2 in Cheremisinoff, N.P. (2002). Handbook of water and wastewater treatment technologies (pp. 372–445). Copyright © 2002 Elsevier Inc.

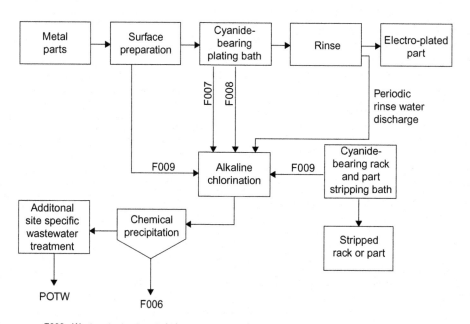

FIGURE 6.6

Hypothetical bearing electroplating process.

From Lewinsky, A. A. (2007). Hazardous materials and wastewater: Treatment, removal and analysis. New York: Nova Science Publishers Inc.

FIGURE 6.7

Thioether-functionalized mesoporous fiber membranes.

From Teng, M., Wang, H., Li, F., & Zhang, B. (2010). Thio-ether-functionalized mesoporous fiber membranes: sol-gel combined electrospun fabrication and their applications for Hg^{2+} removal. Journal of Colloid and Interface Science, 23–28.

(PVP)/SiO_2 nanofibers to make the final product. These mesoporous fiber membranes have a very high selective adsorption of Hg^{2+} and are very easy to recover from a large sample of, say for example, wastewater. This makes them the perfect candidate for this specific heavy ion removal. Fig. 6.7 shows a magnified view of the thioether membranes.

The Hg^{2+} adsorption capacity of the thioether-functionalized (PVP)/SiO_2 membranes was maintained after being recycled three times. It is this combined with the ease of recovery of the membranes that makes them very promising for heavy metal ion removal (Teng et al., 2010).

6.3.3 SOLAR ENERGY

6.3.3.1 Introduction

Renewable energy is becoming an increasingly important topic with depleting fossil fuels and the high rates of pollution. Solar energy is an alternative energy source that involves harnessing the radiant light energy emitted by the sun and converting it into electrical current. Since the middle of the 20th century, the ability to harness and utilize solar energy has greatly increased, making it possible for homes and businesses to make use of the renewal energy source rather than rely on more conventional means of generating power. Research into the applications of solar energy continues, along with the development of more cost-effective ways to capture and store the energy for future use.

At present, the most common means of harnessing solar power is the utilization of a system involving a series of solar panels and storage batteries. The panels collect the radiant light and store the captured energy in the batteries. While energy is stored, it can also be used real time to operate various types of machinery and home appliances. The excess is stored for use at night or in other situations where radiant light is not readily available for some reason.

6.3.3.2 Photovoltaic cells

Photovoltaics is the method of direct conversion of light into electricity at the atomic level. Some materials exhibit a property known as the photoelectric effect that causes them to absorb photons of light and release electrons. When these free electrons are captured, it results in an electric current that can be used as electricity.

Solar cells are made of the same kinds of semiconductor materials, such as silicon, used in the microelectronics industry. For solar cells, a thin semiconductor wafer is specially treated to form an electric field, positive on one side and negative on the other. When light energy strikes the solar cell, electrons are knocked loose from the atoms in the semiconductor material. If electrical conductors are attached to the positive and negative sides, forming an electrical circuit, the electrons can be captured in the form of an electric current—that is, electricity. This electricity can then be used to power a load, such as a light or a tool.

A number of solar cells electrically connected to each other and mounted in a support structure or frame is called a photovoltaic module. Modules are designed to supply electricity at a certain voltage, such as a common 12 V system. The current produced is directly dependent on how much light strikes the module.

Multiple modules can be wired together to form an array. In general, with a larger area of a module or array, more electricity can be produced. Photovoltaic modules and arrays produce DC electricity. They can be connected in both series and parallel electrical arrangements to produce any required voltage and current combination.

Roy, Shastri, Imamuddin, Mukhopadhyay, and Bhasker (2010) explored the application of nanostructured carbon and polymer materials for use in solar energy conversion devices. Fullerene doped with endohedral germanium (Ge) and gadolinium (Gd) and functionalized doped fullerene were mixed with region-regular poly(3-hexyl thiophene) (rr-P3HT), to form an ITO/rr-P3HT-fullerene (Ge/Gd) bulk hetero junction solar device. A decrease in resistance was detected when the temperature of the rr-P3HT compound was increased, and since the temperature of the solar energy material will undoubtedly increase when exposed to sunlight, an adequate temperature range was found. This temperature relationship can be seen in Fig. 6.8.

It is clear that by choosing an adequate functional group, the physical properties of rr-P3HT can change. An increase in UV absorption could potentially lead to more efficient photovoltaic applications.

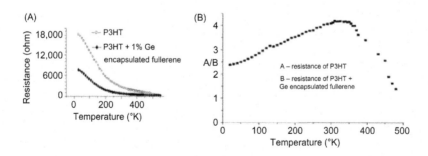

FIGURE 6.8

Temperature dependence of rr-P3HT and rr-P3HT with 1% by weight Ge. (A) Change in resistance as a function of temperature for rr-P3HT and ITO/rr-P3HT-Fullerene. (B) Overall change in resistance ratio as a function of temperature.

From Roy, D., Shastri, B., Imamuddin, M., Mukhopadhyay, K., & Bhasker, K. U. (2010). Nanostructured carbon and polymer materials—synthesis and their applications in energy conversion devices. Renewable Energy, 1014–1018.

We cannot discount the development of strategies to organize ordered assemblies of two components on electrode surfaces, which will lead to a performance improvement of nanostructured organic solar cells (Roy et al., 2010).

6.4 **BIOMEDICAL TECHNOLOGY**

Biomedical technology is a large field of nanotechnology research. It usually involves drug delivery, imaging technology, tissue engineering, and other applications. Tissue engineering will be the focus of the biomedical applications since polymeric nanofibers have a large application in tissue engineering. In tissue engineering, biomaterials are the most important area of research since tissue engineering involves synthetic material dealing directly with the biological material. A brief introduction to wound healing will be examined followed by biomaterials and how they apply to tissue engineering.

6.4.1 **WOUND HEALING**

The wound healing process is important to examine when dealing with biomaterials and surgical implants even at the nanoscale. The process occurs in several stages: (1) hemostasis, (2) inflammation, (3) proliferation, and (4) remodeling. Hemostasis is a relatively simple step where platelets in the blood aggregate and coagulate, resulting in clotting and the cessation of active bleeding. The next stages require a further examination and will be presented in more detail.

6.4.1.1 *Inflammation*

Inflammatory factors are released when collagen comes in contact with blood, by platelets. These factors cause the blood vessels to dilate and become porous: vasodilation. This occurs after a stage of vasoconstriction where the blood vessel constricts to restrict blood flow after the cell membrane is breached. Polymorphous neutrophils, which are attracted by growth factors released after platelets aggregate, proceed to clean the wound and get rid of debris. Macrophages then become the predominant cells, after polymorphous neutrophils. The macrophages destroy damaged tissue and bacteria and release factors that induce angiogenesis, the creation of blood vessels, and stimulate epithelial cells to perform its duties. It is important to note that if the inflammation phase lasts too long, chronic wounds begin to develop which can take years to heal. Inflammation then stops and the proliferation phase begins.

6.4.1.2 *Proliferation phase*

The beginning of the proliferation phase is marked by fibroblasts beginning to enter the wound site. Fibroblasts are responsible for laying down the collagen matrix in the wound site and accumulate at the site simultaneously with angiogenesis. Fibroblasts will initially use the fibrin cross-linking fibers that are formed at the end of the inflammation phase and eventually deposit collagen. Angiogenesis occurs by fibroblast proliferation when endothelial cells move to the wound. Epithelialization then occurs to create epithelial cells in an upward motion. If the basement membrane, the membrane separating the epithelial cells from the collagen, is broken, this process must occur from the margins and skin appendages.

6.4.1.3 *Maturation and remodeling*

When the levels of collagen production and degradation equalize, the maturation process is said to have begun. The strength of the tissue gradually begins increasing.

6.4.2 BIOMATERIALS

Biomaterials are materials that are implanted as prostheses in the body. Prostheses have been used for decades. In the 1920s and 1930s, materials such as vanadium steels, stainless steel, cobalt, titanium, and gold have been used. These materials lacked the surface properties that would make them biologically compatible.

Although strength in the material can be given by the bulk material, the surface properties have an important role since they need to directly interact with the body. For obvious reasons, the surface material must not be toxic or carcinogenic in any way. It must also be biocompatible, being recognized as non-foreign material. If the material is recognized as a foreign material, the body will not accept it and proper healing will not occur, as described in the wound healing section discussed previously. Table 6.2 displays common materials used in medicine currently.

Biocompatibility is a difficult term to clearly define, but it is often defined in terms of if the material fulfils its intended purpose. This may have different characteristics depending on what type of application it is used for. Some applications require adhesion to cells. Biocompatible also does not mean being without side effects. There may be minimal side effects even though the material may be performing its duty well.

Polymeric nanofibers have the application of creating an extracellular matrix (ECM) that is often needed in the wound healing process. This has the potential to speed the healing process if the material is biocompatible. Natural polymers have also been integrated in the nanofiber. Lecithin and collagen have been known to increase the biocompatibility and even allow for the cells to proliferate easier on the ECM. This is due to the presence of natural proteins in the nanofiber, leading to a higher cytocompatibility. The presence of a foreign body can lead to the host cells rejecting the implant in a process called foreign body reaction.

Table 6.2 Applications of Synthetic and Modified Biological Materials

Material	Application	Tissue Response
Titanium and alloys	Joint prostheses, oral implants, fixation plates, pacemakers, heart valves	Inert
CaP ceramic	Joint prostheses, oral implants, bone replacement, middle ear replacement	Bioactive
Alumina	Joint prostheses, oral implants	Inert
Carbon	Heart valves	Inert
Poly(tetrafluoroethylene)	Joint prostheses, tendon and ligament replacement, artificial blood vessels, heart valves	Inert
Poly(methylmethacrylate)	Eyes lenses, bone cement	Tolerant
Poly(dimethylsiloxane)	Breast prostheses, catheters, facial reconstruction, tympanic tubes	Unknown
Poly(urethane)	Breast prostheses, artificial blood vessels, skin replacements	Inert
Poly(lactic) acid	Bone fixation plates, bone screws	Inert
Poly(glycolic) acid	Sutures, tissue membranes	Inert

From Malsch, N. H. (2005). Biomedical nanotechnology. Boca Raton: Taylor & Francis Group.

6.4.2.1 *Foreign body reaction*

There are four potential results of a foreign body being introduced in a biological system: (1) integration, (2) extrusion, (3) resorption, or (4) encapsulation. Integration is obviously the most desired outcome of a biomaterial implantation, since it will interact with the biological system as if it were a part of the system. The number of cases of true biointegration achieved has been limited, most frequently they have been cases of bone tissue implants with titanium coated with hydroxyapatite. Most often soft tissue, such as skin, implantation results in the other three outcomes. Extrusion occurs when the implanted device is in contact with epithelial cells and the epithelium will form a pocket around the implantation. If it is close to the surface of the epithelium, it will be gradually pushed out of the host. Resorption occurs when the implant is made of degradable material and the implant degrades, resulting in a collapsed scar at the implantation site. Encapsulation of soft tissue is the most frequent foreign body reaction. The capsule consists of a membrane of a high amount of collagen with a layer of myofibroblasts outside the membrane. The implant is then isolated from the body (Malsch, 2005).

Nirmala et al. (2010) attempted to create a blend of lecithin and polyamide-6 nanofibers that would allow osteoblast cell cultures to attach to and grow on the nanofiber blend. Polyamide-6 with 0, 1, 3, 5 wt% lecithin solution was used to prepare the nanofiber composite. Different instruments were used to characterize the nanofibers including average pore diameter, pore volume, pore area, total porosity, structure by X-ray diffraction, and bonding configurations with Fourier-transform infrared. Osteoblast cell culture was introduced after carefully preparing and incubating the cultures (with fetal calf serum). The osteoblast cells were grown on the polymer scaffolds and cultured.

All nanofibers were comprised of randomly oriented fibers with smooth surfaces and uniform diameters. The polyamide-6 nanofibers created 150–200 nm structures whereas the mesh-like nanofiber structure resulted in one magnitude less than that (10–30 nm). The porosity is important to provide framework for the seeded cells. The 5% lecithin blend provided higher porosity and diameter with much lower average pore volume than polyamide-6 unblended. There is improved water wettability for the higher wt% lecithin than pristine polyamide-6. Wettability is advantageous since osteoblastic culture can spread more easily on hydrophilic surfaces, rather than hydrophobic ones. The lecithin blends showed a melting peak at 220–215°C and the crystallinities of the nanofibers were ~185–180°C.

Fig. 6.9 shows the cell culture after 3 days on the various nanofiber blends after adhesion and proliferation were examined. The cells are clearly growing attached to the matrix. The cytotoxic effect of the nanofiber was measured by measuring levels of pyruvic acid with a spectrophotometer. The lactate dehydrogenase activity in the medium was examined. There was a slight increase in toxicity as the lecithin was incorporated into the cells.

It was concluded that the polyamide-6/lecithin composite nanofibers improve the cell growth behaviors. If the mechanical strength of the scaffold can be increased, the nanofiber can be used practically for bone regeneration. It can be used in vitro for transplantation to the site.

By examining the biocompatibility of human umbilical vein endothelial cells (HUVECs) on a poly(L-lactide-co-ε-caprolactone) (PLCL)/fibrinogen nanofiber blend, a biodegradable ECM capable of supporting some types of tissues can be created. PLCL is a biodegradable polymer which shows promise in soft tissue engineering, and fibrinogen is a soluble plasma glycoprotein, used to enhance the adhesion of cells to PLCL. Using the random copolymer PLCL with 70 mol% L-lactide, an 8% PLCL solution was prepared. Similarly, a fibrinogen solution was prepared and they were mixed at 4:1,

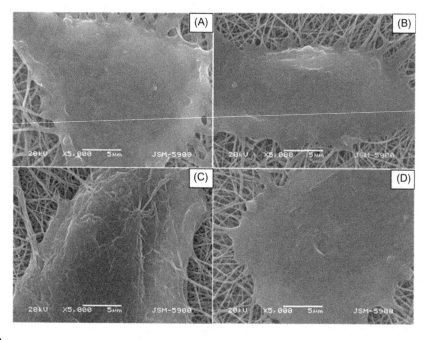

FIGURE 6.9

SEM image of cell growth on nanofiber with lecithin concentrations of (A) 0, (B) 1, (C) 3, and (D) 5wt%.

From Nirmala, R., Park, H. M., Navamathavan, R., Kang, H. S., El-Newehy, M., & Kim, K. Y. (2010). Lecithin blended polyamide-6 high aspect ratio nanofiber scaffolds via electrospinning for human osteoblast cell culture. Material Science and Engineering C, 486–493.

2:1, 1:1, 1:2, pure PLCL, fibrinogen, and tissue culture polystyrene as a control. This was put through an electrospinning process. The different composites were placed in culture plates with HUVEC. Different tests were performed including cell proliferation tests, cell adhesion tests, and morphological characteristic tests.

It was found that the higher the fibrinogen content in the composite, the smaller the water contact and the smaller the fiber diameter of the nanofiber. It was found that initially there was no significant difference between the different composites, but after 3–5 days, the PLCL/fibrinogen blended scaffolding had increased the cell number quickly especially compared to the pure PLCL nanofiber. After day 7, it was found that the proliferation rates of the PLCL/fibrinogen nanofibers were 3.23, 4.35, 3.95, and 3.39 for 4:1, 2:1, 1:1, and 1:2 ratios, respectively. Fibrinogen had a much higher proliferation rate, but this had been previously known. Fig. 6.10 shows the proliferation of cells on the nanofiber scaffolds by way of metabolic activity of the cells. The PLCL/fibrinogen blend of 2:1 may be the most suitable culture. The live/dead assay demonstrated that very few cells were found dead possibly demonstrating that the nanofibers were relatively non-cytotoxic. This may be a good sign for future applications (Fang et al., 2010).

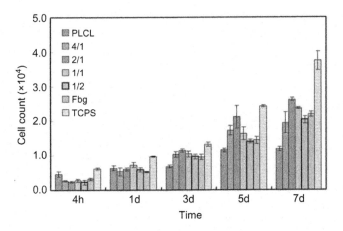

FIGURE 6.10

Metabolic activity of cultured HUVECs on PLCL/fibrinogen nanofiber scaffolds.

From Fang, Z., Fu, W., Dong, Z. D., Zhang, X., Gao, B., Guo, D., et al. (2010). Preparation and biocompatibility of electrospun poly (L-lactide-co-ε-caprolactone)/fibrinogen blended nanofibrous scaffolds. Applied Surface Science, 4133–4138.

Nguyen, Kin, Song, and Lee (2010) examined the process of thermally treating Ag nanoparticles (NPs) within polyvinyl alcohol (PVA) nanofibers as a means to inhibit microbial growth for skin applications. The silver interrupts microbial growth that can hinder the process of skin regeneration.

An aqueous solution of $AgNO_3$ was added to distilled water and irradiated with microwaves, reducing the Ag ion to have no charge. The PVA and Ag solution was then electrospun and heated from 0°C to 150°C at 5°C/min in a nitrogen atmosphere. A test of inhibition with *Staphylococcus aureus* and *Escherichia coli*, gram-positive and gram-negative bacteria, respectively, was carried out in an incubator at 37°C for 18–24 h.

It was first found that the increased microwave irradiation time increased the diameter of the nanofibers, 100–200 nm and 100–500 nm for the 60 s and 90 s irradiation, respectively. This was due to the evaporation of water increasing the polymer concentration in solution. The samples vary in Ag concentration with sample A having a 9.93% Ag concentration and sample B with 5.73%. Sample A was irradiated for 60 s while sample B was irradiated for 90 s. Using UV-Vis absorption, it was confirmed that Ag NPs were formed in the nanofibers. Fig. 6.11 shows the SEM imaging of the nanofibers loaded with Ag NPs. The electrospun mats were then heated for 24 h to 80°C, 120°C, and 150°C. The heat treatment at higher temperatures was shown to have Ag NPs that were larger in size on the surface of the nanofiber. It was found that the higher-temperature-treated nanofibers had a slightly increased antimicrobial property. The nanocomposites were much stronger and more brittle than pure PVA mat, with the heat treatment having an improvement on the strength of the nanofiber. The nanofibers showed antibacterial properties in a zone around the application area with the 60 s irritation sample at 150°C heat treatment having the best antimicrobial properties. The Ag NP loaded PVA shows great promise in the area of skin application to promote rapid healing (Nguyen et al., 2010).

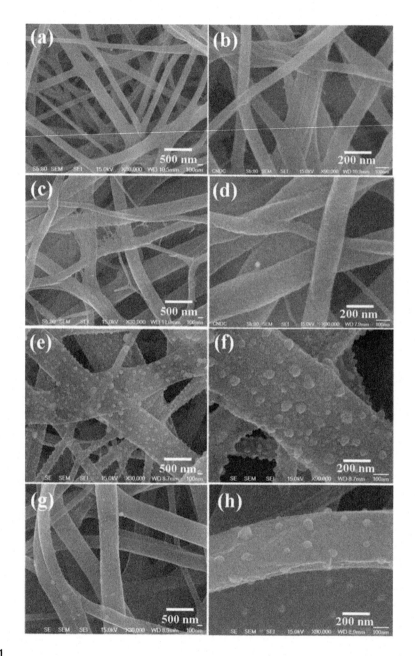

FIGURE 6.11

SEM micrograph of Ag NP loaded PVA nanofiber mats with 60s irradiation and heat treatment at (A,B) 80°C, (C,D) 120°C, (E,F) 150°C and with 90s irradiation and heat treatment at (G,H) 150°C.

From Nguyen, T.-H., Kin, Y.-H., Song, H.-Y., & Lee, B.-T. (2010). Nano Ag loaded PVA nano-fibrous mats for skin application. Journal of Biomedical Materials Research B, 225–233.

6.5 **POLYMERIC NANOFIBERS AS SENSORS**

In this section, NPs used for sensory applications are presented; they are known as nanosensors. This section will outline the classification of nanosensors, their applications, and how they operate.

6.5.1 **CLASSIFICATION OF POLYMERIC NANOFIBERS**

Nanosensors are sensors on the "Nano" scale. "Nano" is a unit of measurement for 10^{-9} m. Fig. 6.12 shows a polymeric nanosensor which has been embedded onto a piece of plastic.

It is remarkable how small the sensor actually is, because there are many sensors placed onto the surface of the plastic, and the sensor is part of an "intelligent transportation system" used to monitor traffic and help reduce congestion. The nanosensor is made from a carbon nanotube (plastic)/cement composite. A very interesting component of the application is that the nanotube/cement sensor composite could be used as pavement, which could also measure traffic flow (Nano-Tech-Views, n.d.).

6.5.2 **HOW A POLYMERIC NANOFIBER SENSOR WORKS**

Nanopolymer sensors utilize electrostatic interactions between molecules to interact with their surroundings. An electric current is formed in the presence of a particular compound. The compound exerts an electrostatic force on the nanosensor which pushes electrons in the sensors, thereby forcing a current. This current is measured and correlated to the concentration of the particular compound being measured. This requires the sensor to be calibrated on a typical electrostatic force, from the usual background. For example, a sensor is placed in an environment to measure the presence of gaseous hazards in air. The sensor must be calibrated to the typical air contents beforehand. When the air contains a sufficient amount of hazardous compounds, the hazardous compounds exert an electrostatic force onto the sensor, creating a current to be detected.

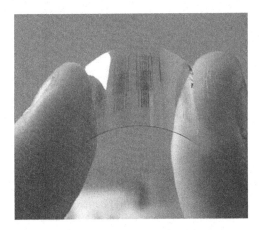

FIGURE 6.12

A nanosensor embedded on a piece of plastic.

From Nano-Tech-Views, Cheap, sensitive Stanford sensors could detect explosives, toxins in water.

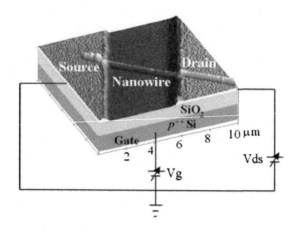

FIGURE 6.13

Schematic of electrical circuit utilizing a nanowire.

From Jia, G. L. (2006, June 8). Electrically controlled nanowire-based chemical sensors. Retrieved 13.04.11, from SPIE: http://spie.org/x8765.xml?ArticleID=x8765

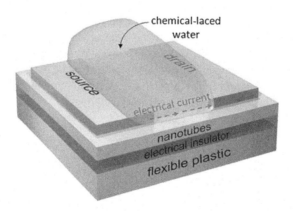

FIGURE 6.14

Schematic of electric current flow in sensory applications.

Courtesy of Mark Roberts, http://www.clemson.edu/ces/departments/chbe/people/roberts_m.htmlpublication.

Fig. 6.13 shows a simplistic circuit using a polymer nanowire (Jia, 2006). Fig. 6.14 shows the direction of an electrical current when a chemically laden solution, able to exert electrostatic forces onto the sensor, is present. The sensing mechanism is from the charge transfer between the semiconducting material (which is doped onto a polymer base) and the chemical species which has been adsorbed at the exposed surface of the sensor. Included in the circuit are:

- the source, where the current comes from (from adsorbed molecules),
- the nanowire, which conducts the current,

- the drain, which gives the direction for current (opposite voltage when compared to the source),
- the ground, which is always needed in a circuit, acting to ensure the safety of the circuit, giving or absorbing extra electrons.

The sensitivity decreases when increasing the gate voltage because the gate potential can electrically desorb the absorbed molecules on the exposed surface of the sensor.

6.5.3 FABRICATION OF NANOPOLYMERIC SENSORS

Fabrication techniques regarding how to make nanopolymers have been discussed. This section specifically discusses how polymeric nanosensors are made.

Polymeric nanosensors are made from techniques such as bottom–up assembly and top–down lithography, each of which are discussed below:

Bottom–up assembly: This technique pertains to creating the nanosensor on the molecular level, assembling each individual molecule in a series to achieve a larger sized product.

The advantages of bottom–up assembly include having few defects created during fabrication. This is because formation occurs on the individual molecule basis, which means every molecule can be positioned where it is desired. An example of bottom–up assembly is electrospinning, which is the primary method of fabricating polymeric nanosensors, as well as various other advanced nanopolymers seen throughout the chapter.

The disadvantages of bottom–up assembly include not being able to easily mass produce the compound. This is because the bottom–up method goes molecule by molecule, which is very time consuming.

Top–down lithography: This technique involves taking a millimeter- or centimeter-sized object and breaking it down into the desired shape of the sensor. This is usually done by short wavelength electromagnetic sources.

The greatest advantage of the process is that there is no assembly process putting the molecules together. This allows mass production to take place very easily.

The disadvantage of this process is that there is a higher probability for defects. This occurs because the procedure is taking a large compound and breaking it down.

Generally, polymeric nanosensors are made by fabricating a polymer base and doping conductive material (usually a semiconductor, but it could include other compounds) in the base. The method of doping can include many processes dependent on the shape of the polymeric base. A very common technique of doping includes chemical vapor deposition.

A polymeric base could have many different shapes. The base could be made of nanofibers, nanotubes, nanospheres, as well as other geometries.

6.5.4 TYPES OF SENSORS

There are many different types of sensors, measuring different things. This includes the following types of sensors:

- *Piezoelectric*—measuring mass
- *Electrochemical*—measuring electric conductivity and distribution (voltage)
- *Optical*—measuring the intensity of light or radiation
- *Calorimetric*—measuring the flux of heat into the sensor.

Polymeric nanosensors can be used in many applications associated with any of the above, which could include almost anything tangible. This can be applied to any industry that needs any type of sensory system.

The nanopolymer sensors have a wide variety of applications, which include measuring the:

- concentration of gaseous compounds in air: oxygen, carbon dioxide, ammonia, and volatile organic compounds, as well as others,
- humidity (water content) in air,
- concentrations of liquids in solution: mercury and various other compounds.

Not only can nanosensors be applied to a variety of processes, but they can replace older macromolecular sensors. This is because nanosensors analyze the concentration of various compounds on the molecular level. This means the nanosensor can measure concentrations on a very small scale, with a significantly higher sensitivity than macromolecular sensors. Nanosensors can accurately measure very small changes in a compound`s concentration.

Sensing makes use of an electric current to generate a signal. Therefore, to have a polymer conduct a signal, it must exhibit a certain amount of polar behavior. This is achieved by doping the polymer with conducting compounds or by using polymers that have moderately polar repeating chains. All of the conducting materials used in the following experiments exhibit some sort of molecular polarity.

A polymer comprised of poly(1,5-diaminonapthlene) (DAN) nanofibers was produced by Rahman et al. (2008) through a chemical catalyst polymerization, using Iron(III) salt as catalyst. Such nanofibers were used to sense water in a nonaqueous solution of acetonitrile.

The sensors were fabricated by recrystallizing the DAN nanofibers twice in an aqueous ethanol (50%v/v) solution under nitrogen and dried at room temperature under vacuum over P_2O_5 for 48 h. Poly(1,5-DAN) nanofibers were fabricated using chemical catalytic polymerization. The fabrication procedure had 2 g of DAN monomer being added to 100 mL methanol solution containing 3.4 g $FeCl_3$. The mixture was stirred at 30°C for 24 h. The reaction was done in both oxygen and nitrogen atmospheres.

Various spectroscopic methods showed that the polymer was formed homogeneously. Fibers were found to be roughly 10–30 nm in diameter and 400 nm in length. Fig. 6.15 shows the experimental results obtained.

The graph on top (A) shows the variance of current with time, this is very important because having a constant current allows the sensors to receive a constant signal, so a proper display of the reading can be made. Graph (B) gives the current and applied voltage for various different acetonitrile solutions containing a water concentration of (starting from the bottom): (a) 1%, (b) 5%, (c) 10%, (d) 20%. The significance of graph (B) is it outlines the current expected for the applied voltage the circuit is operating under. Finally, graph (C) gives the expected current generated at various temperatures, the percentages in graph (C) correlate to the water concentration in solution. Note that the highest concentration of water in solution gives the highest current, this is correlated to the fact water is a conductor, and having a greater presence of water in solution gives a greater amount of current during operation.

The concentration of $FeCl_3$ and the solvent used was shown to have a significant effect on the reaction yield. Using methanol solvent also produced a very high reaction yield. The DAN sensor is very sensitive to water in a nonaqueous acrylonitrile solution. The detection limit of the sensor was found to be 0.01%.

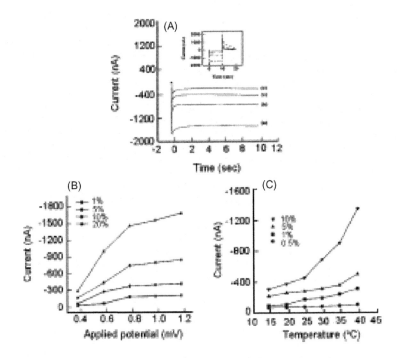

FIGURE 6.15

Results of experimentation. (A) Variance of current with time in acrylonitrile solution containing water
at (a) 20%, (b) 10%, (c) 5%, (d) 1% concentrations. (B) Current and applied voltage relationship.
(C) Temperature dependence on current.

From Rahman, A., Won, M.-S., Kwon, N.-H., Yoon, J.-H., Park, D.-S., & Shim, Y.-B. (2008). Water sensor for a nonaqueous solvent
with poly(1,5-diaminonapthalene) nanofibers. Department of Chemistry and Center for Innovative BioPhysio Sensor Technology.

The results were comparable to the Karl Fisher titration method. The Karl Fisher titration is problematic because it is a time-consuming process, and it is difficult to determine the endpoint of the titration.

In a different study, Aussawasathien, Dong, and Dia investigated two different uses of polymeric nanofiber sensors: one is to sense the humidity, while the other is used to sense the presence of hydrogen peroxide and glucose.

The nanofibers were prepared via electrospinning and had a diameter of between 400 and 1000 nm. These sensors had significantly more sensitivity compared to their film-type counterparts. It was noticed that after use, the sensor measuring humidity had small morphological changes, while the sensor measuring hydrogen peroxide and glucose exhibited no changes.

They made a PEO/LiClO$_4$ sensor that was used with a film-type sensor to measure humidity in a humidifier at 25°C at a range of 25–65% humidity. They also made a PA/PS/CSA sensor containing GOx that was used with a film-type sensor to measure H$_2$O$_2$ from the oxidation of glucose at hydrogen peroxide concentrations less than 25 mM.

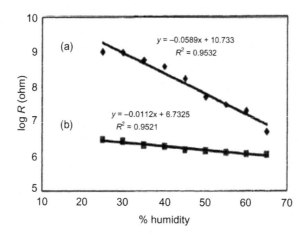

FIGURE 6.16

Results for PEO/LiCIO$_4$ humidity sensor compared to the film sensor. (a) The results for the PEO/LiCIO4 sensor and (b) the film sensor. Log "R" is the logarithm of resistivity and "% humidity" is the percentage humidity.

From Aussawasathien, D., Dong, J.-H., & Dia, L. (n.d.). Electrospun polymer nanofiber sensors. Department of Materials and Chemical Engineering and University of Dayton Research Institute.

The results of the experiment gave the following graphs: the first pertains to the PEO/LiCIO$_4$ sensor while the second pertains to the PA/PS/CSA sensor (Fig. 6.16).

Fig. 6.16 shows the change in resistivity with percentage humidity. The greater slope above shows the greater sensitivity of the nanopolymeric sensor compared to film sensors.

Fig. 6.17 shows the varying current with hydrogen peroxide concentration. It also shows that the nanopolymeric sensor is significantly more sensitive than the film sensor.

It was found that the polymeric nanosensor changes the current transmitted very quickly. This means that the nanosensor is more sensitive than the film sensor. A greater sensitivity is desired because it distinguishes the current conditions better than a less sensitive device. The sensitivity is improved by the contact surface area of the sensor.

Lee, Oh, Kang, and Kwak (2010) presented sensors that sense nitroaromatic compounds including: 2,4-dinitrotoluene and 2,6-dinitrotoluene. They also note that similar types of sensors may be applied to sense volatile organics. The polymers used were: poly[1-phenyl-2-(p-trimethylsilyl)phenylacetylene] (PTMSPA) and poly[1-phenyl-2-p-(dimethyl octadecylsilylphenyl)acetylene] (PDMOSPA). Both of these polymers have high molecular weight and high polydispersity indices.

Polymer solutions were prepared by mixing benzene with one of the above mentioned polymers (the solution created is 3 mL) and put into a vail. The vail was deep-freezed in a deep freezer (−70°C) or with liquid nitrogen (−196°C). After freezing the sample was dried in a freeze dryer (−50°C at a pressure of 9 mmHg).

The photoluminescence spectra were analyzed on a spectrofluorometer. The fluorescence of PTMSPA was found by inserting the fibers into a vail containing the solid analytes (2,4-dinitrotoluene or 2,6-dinitrotoluene); this was at room temperature, with an excitation wavelength of 420 nm. The structural coarsening was analyzed along two routes: first by freezing the solution and second

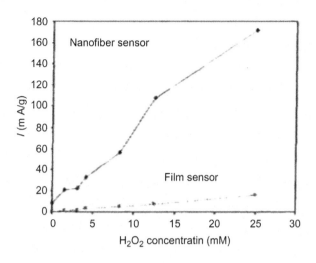

FIGURE 6.17

Results for PA/PS/CSA hydrogen peroxide sensor. The vertical axis is milliamps/gram of PA/PS/CSA polymer and the horizontal axis is the concentration of H_2O_2 in mM.

From Aussawasathien, D., Dong, J.-H., & Dia, L. (n.d.). Electrospun polymer nanofiber sensors. Department of Materials and Chemical Engineering and University of Dayton Research Institute.

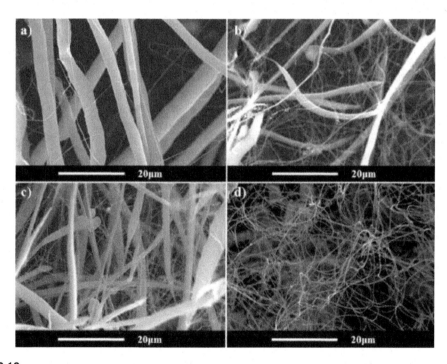

FIGURE 6.18

Picture of nanofibers at various mass fractions: (A) 0.1%, (B) 0.01%, (C) 0.001%, and (D) 0.0003% wt.

From Lee, W.-E., Oh, C.-J., Kang, I.-K., & Kwak, G. (2010). Diphenylacetylene polymer nanofiber mats fabricated by freeze drying: preparation and application for explosive sensors. Macromolecular Journals.

by the sublimation of the frozen solution in a vacuum. The PTMSPA polymers were fabricated with four different mass fractions (in benzene): 0.1%, 0.01%, 0.001%, and 0.0003%. These different mass fractions had their properties analyzed after freeze drying in a deep freezer, and after they were frozen using liquid nitrogen.

PTMSPA and PDMOSPA polymer solutions were successfully prepared with a large fractional free volume. It was found that the morphology of the structure of the fibers are significantly dependent on the polymer solution, the composition of the polymer, and the freezing methodology (either using liquid nitrogen or a freeze drier). Fig. 6.18 shows the various thicknesses found when utilizing different mass fractions of polymer in solution.

These studies show the very interesting potential for nanofiber polymers to act as sensors. These sensors can accurately sense various compounds in the environment, such as: the concentration of water in nonaqueous liquids, the presence of nitroaromatic compounds in vapor, and the presence of hydrogen peroxide and glucose.

6.6 CONCLUSIONS

There are many different applications for polymeric nanofibers, and most of these are still being researched and developed. Biomedical technology has a diverse set of tools to further medical sciences. Tissue engineering is a complex but interesting application, using nanotechnology to interact closely with cells. The role of polymeric nanosensors can revolutionize the sensory industry, with applications ranging from measuring the flow of traffic to the concentration of hazardous gases in the air. Finally, regarding the three environmental applications discussed in this section, nanofibers have significantly improved the efficiency and practicality of each. With developments in mass production, nanofibers seem to be on the road to becoming the standard material to use in manufacturing various fabrications. Environmental issues we face every day cannot be solved overnight, but with the increasing discoveries in nanotechnology we will achieve this goal significantly sooner. In conclusion, the role of nanopolymers in human society is very significant and will be utilized more in the future.

NANOTUBES*

7.1 INTRODUCTION

Carbon nanotubes (CNTs) are one of the most important materials under investigation for nanotechnology. Their tensile strength is one hundred times that of steel, their thermal conductivity is found to be better than all but the purest diamond, and their electrical conductivity is similar to copper but with the ability to carry much higher currents (Holister, Harper, & Vas, 2003). NTs come in different shapes: long, short, single walled, multi-walled, open, closed, with different types of spiral structure, etc. Each type has its own production costs and applications. Some have been produced in large quantities while others are only now being produced commercially with average purity and in quantities greater than a few grams. NTs have been constructed with length to diameter ratio of up to 132,000,000:1, which is much larger than any other material. They are considered as members of the fullerene structural family, which also includes the spherical buckyballs. Their name is derived from their size, since the diameter of an NT is on the order of only a few nanometers which is approximately 1/50,000th of the width of a human hair, while their length can be up to 18 cm. NTs are categorized as single-walled nanotubes (SWNTs) and multi-walled nanotubes (MWNTs) (Holister et al., 2003). Fig. 7.1 shows the two types of NTs.

The term "nanotube" is usually used to refer to the CNT, which has received high attention from researchers over the last few years and promises a host of interesting applications. There are many other types of NTs, from various inorganic kinds, such as those made from boron nitride, to organic ones, such as those made from self-assembling cyclic peptides or from naturally occurring heat shock proteins that are extracted from bacteria that thrive in extreme environments. However, CNTs excite the most interest, promise the greatest variety of applications, and currently appear to have by far the highest industrial potential (Holister et al., 2003).

7.1.1 HISTORY

In 1952, Radushkevich and Lukyanovich published clear images of 50 nm diameter tubes made of carbon in the *Soviet Journal of Physical Chemistry*. Later, a paper by Oberlin, Endo, and Koyama published in 1976 showed hollow carbon tubes with nanometer-sized diameters using a vapor growth technique. In 1979, John Abrahamson presented evidence of CNTs at the 14th Biennial Conference of Carbon at Penn State University. The paper described how CNTs were produced on carbon anodes

---

* By Yaser Dahman, Ahmad Bayan, Bohdan Volynets, and Navid Ghaffari.

Nanotechnology and Functional Materials for Engineers. DOI: http://dx.doi.org/10.1016/B978-0-323-51256-5.00007-1

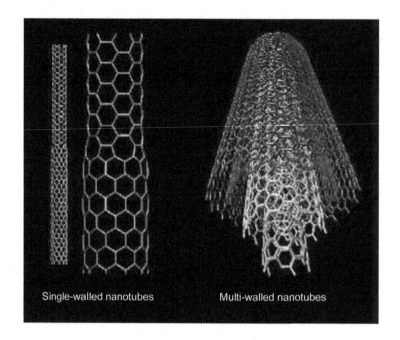

FIGURE 7.1

Single-walled and multi-walled nanotubes.

From Holister, P., Harper, T. E., & Vas, C. R. (2003). Nanotubes white paper. CMP Cientifica, 5–7.

using the arc discharge apparatus. Later on, in 1981, a group of Soviet scientists published papers about carbon nanoparticles produced using the thermocatalytical disproportionation of carbon monoxide method. The authors mentioned in their paper that the described carbon multilayer tubular crystals were created by rolling graphene layers into cylinders. Most of the academic literature credits the discovery of hollow, nanometer-sized tubes composed of graphitic carbon to Sumio Iijima of Nippon Electric Company in 1991 (Hirlekar et al., 2009).

7.1.2 CLASSIFICATION OF CNTs

CNTs are classified into the following two types: SWNT and MWNT.

Comparison between SWNT and MWNT is presented in Table 7.1.

7.1.3 MOLECULAR STRUCTURE

A CNT can be described as a cylindrical molecule composed of carbon atoms. A typical single-walled carbon nanotube (SWCNT) structure is shown in Fig. 7.2. A major feature of the structure is the hexagon pattern that repeats itself periodically in space. As a result of the recurrence, each atom is bonded

Table 7.1 Comparison Between SWNT and MWNT

SWNT	MWNT
Single layer of graphene	Multiple layer of graphene
Catalyst is required for synthesis	Can be produced without catalyst
Bulk synthesis is difficult as it requires proper control over growth and atmospheric condition	Bulk synthesis is easy
Purity is poor	Purity is high
A chance of defect is more during functionalization	A chance of defect is less but once occurred it is difficult to improve
Less accumulation in body	More accumulation in body
Characterization and evaluation is easy	It has very complex structure
It can be easily twisted and is more pliable	It cannot be easily twisted

From Hirlekar, R., et al. (2009). Carbon nanotubes and its applications: a review. Asian Journal of Pharmaceutical and Clinical Research, 1–11.

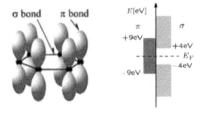

FIGURE 7.2

Scheme of sp^2 hybridization in graphene; σ bonds, π bonds, and their energies with respect to Fermi level.

From Ruoff, R., et al. (2003). Mechanical properties of carbon nanotubes: theoretical predictions and experimental measurements.
C. R. Physique, 4, 993–1008.

to three neighboring atoms. The resulting structure is mainly due to the sp^2 hybridization bonding during which one s-orbital and two p-orbitals combine to form three hybrid sp^2 orbitals at an angle of 120° to each other within the same plane. This covalent bond which is referred to as the σ bond and is shown in Fig. 7.3 plays an important role in the remarkable mechanical properties of CNTs as a result of this strong chemical bond. Also, the π-bond is out of the plane and is relatively weak which contributes to the interaction between the layers in multi-walled carbon nanotubes (MWCNTs) and between SWCNTs in SWCNT bundles (Ruoff et al., 2003).

7.1.4 STRUCTURES OF SWCNTs

The bonding in the CNTs is similar to the graphene sheet. A widely used approach to identify the types of SWCNT is by reference to rolling up the graphene sheet. The key geometric parameter associated

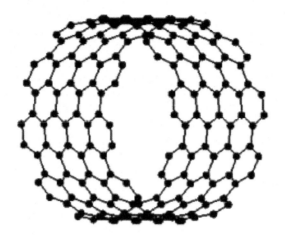

FIGURE 7.3

Structure of a section of (10, 10) CNT. Each node shown is a carbon atom and lines represent the chemical bonds.

From Ruoff, R., et al. (2003). Mechanical properties of carbon nanotubes: theoretical predictions and experimental measurements.
C. R. Physique, 4, 993–1008.

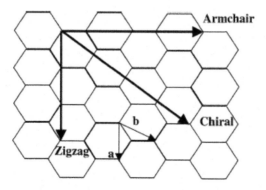

FIGURE 7.4

Definition of roll-up vector as linear combinations of base vectors **a** and **b**.

From Ruoff, R., et al. (2003). Mechanical properties of carbon nanotubes: theoretical predictions and experimental measurements.
C. R. Physique, 4, 993–1008.

with this process is the roll-up vector **r** as shown in Fig. 7.4 which is made to be equal to $n\mathbf{a} + m\mathbf{b}$, and therefore can be expressed as the linear combination of the lattice basis **a** and **b**. It is then possible to associate a particular integer pair (n,m) with each SWCNT. The values of n and m define three categories of CNTs, if: $m=0$, "Zigzag"; $n = m$, "Armchair"; other, "Chiral" (Ruoff et al., 2003). Fig. 7.5 shows the different types of NTs.

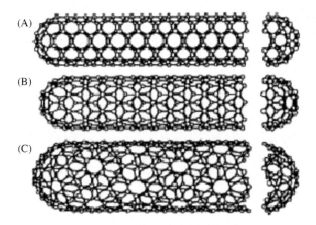

FIGURE 7.5

Different types of NTs. (A) zigzag, (B) armchair, and (C) chiral.

From Ávila, A.F., Lacerda, G.S.R. (2008). Molecular mechanics applied to single-walled carbon nanotubes. Material Research 11(3).
São Carlos July/Sept. 2008.

7.2 PROPERTIES

7.2.1 MECHANICAL PROPERTIES

CNTs are the strongest and stiffest materials created so far. This strength is a result of the sp^2 covalent bonds created between the individual carbon atoms. MWCNTs were found to have a tensile strength of around 63 GPa (Min-Feng et al., 2000). Since CNTs have a low density of around $1.4\,g/cm^3$, their specific strength of up to 48,000 kN m/kg is the best compared to any other material (Collins, 2000).

Under high tensile strain, the NTs will undergo plastic deformation. This deformation begins at strains of around 5% and can increase the maximum strain the tubes undergo before fracture by releasing strain energy (Collins, 2000).

CNTs are not as strong under compression because they are hollow structures with high aspect ratio and they tend to start buckling when stress is forced upon the tubes (Jensen et al., 2007).

Table 7.2 shows the mechanical properties for the different types of NTs.

7.2.2 ELECTRICAL PROPERTIES

The symmetry and the unique electronic structure of graphene affect the electrical properties of NTs and make some of them have semiconducting properties. It was found that for a given (n,m) NT, if $n = m$ or if $(n-m)/3 =$ integer, the NT is metallic; if $(n-m)/3$ is not an integer, then the NT is semiconducting.

On the other hand, MWCNTs that comprise of interconnected inner shells appear to have superconductivity with a relatively high transition temperature $T_c=12\,K$. In comparison, the T_c value is an order of magnitude lower for SWCNTs or for MWNTs with noninterconnected tubes (Popov et al., 2002).

7.2.3 THERMAL PROPERTIES

CNTs are found to be great thermal conductors, displaying a property known as ballistic conduction. It was found that SWNTs have a room-temperature thermal conductivity along their axis of about 3500 W/(m·K)

Table 7.2 Mechanical Properties of NTs

Material	Young's Modulus (TPa)	Tensile Strength (GPa)	Elongation at Break (%)
SWNT	~1 (from 1 to 5)	13–53	16
Armchair SWNT	0.94	126.2	23.1
Zigzag SWNT	0.94	94.5	15.6–17.5
Chiral SWNT	0.92		
MWNT	0.2–0.8–0.95	11–63–150	
Stainless steel	0.186–0.214	0.38–1.55	15–50
Kevlar	0.06–0.18	3.6–3.8	~2

From Bellucci, S. (2005). Carbon nanotubes: physics and applications. Physica Status Solidi, 34–37.

(Pop et al., 2005); comparing this value to copper, conductivity transmits 385 W/(m·K). In addition, it was found that SWCNTs have a room-temperature thermal conductivity across their axis in the radial direction of about 1.52 W/(m·K) (Saion et al., 2005), which is about as thermally conductive as soil.

7.3 PRODUCTION METHODS OF NTs

CNTs are generally produced by three main techniques: arc discharge, laser ablation, and chemical vapor deposition (CVD) (Hirlekar et al., 2009). In the arc discharge method, a vapor is created by an arc discharge between two carbon electrodes with or without a catalyst. NTs self-assemble from the resulting carbon vapor. In the laser ablation technique, a high-power laser beam is directed at a volume of carbon-containing feedstock gas such as methane or carbon monoxide. Currently, laser ablation produces a small amount of clean NTs, whereas arc discharge methods generally produce large quantities of impure material. In general, CVD results in MWNTs or poor quality SWNTs. The SWNTs produced with CVD have a large diameter range, which can be poorly controlled (Hirlekar et al., 2009). On the other hand, this method is easy to scale up, which favors commercial production.

7.3.1 ARC DISCHARGE METHOD

In this method, as shown in Fig. 7.6, NTs are produced through arc vaporization when two carbon electrodes are placed end to end with a distance of about 1 mm in the presence of an inert gas such as helium or argon at a pressure between 50 and 700 mbar. Carbon rods are evaporated when a current of 50–100 A driven by 20 V that generates high temperature is discharged between the two electrodes. Due to this, the anode will evaporate and CNTs will self-assemble on the cathode. Large-scale production of CNTs depends on uniformity of plasma arc and temperature of deposition (Hirlekar et al., 2009).

7.3.2 LASER ABLATION METHOD

In this method, a continuous or pulsed laser is used to vaporize a graphite target in an oven at 1200°C and in the presence of helium or argon gas in order to keep the pressure at 500 Torr. This method is very expensive so it is mainly used for SWNTs. Laser vaporization results in higher yield of SWNTs with narrower size distribution than those produced in the arc discharge process (Hirlekar et al., 2009) (Fig. 7.7).

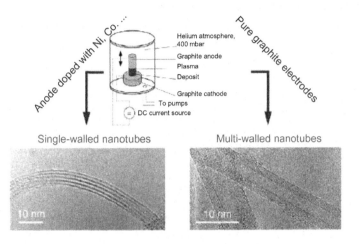

FIGURE 7.6

Experimental setup of an arc discharge apparatus.

Reproduced from Nessim, G.D. (2010). Properties, synthesis, and growth mechanisms of carbon nanotubes with special focus on thermal chemical vapor deposition. Nanoscale 2, 1306–1323, with permission of The Royal Society of Chemistry.

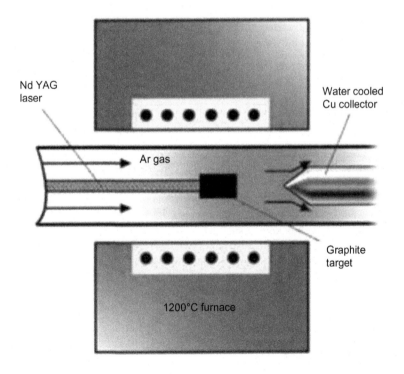

FIGURE 7.7

Schematic drawing of a laser ablation apparatus.

From Yadav, B.C., Kumar, R., Srivastava, R., Shukla, T. (2011). Flame synthesis of carbon nanotubes using camphor and its characterization. International Journal of Green Nanotechnology 3(3), 170–179, Copyright © 2011 Taylor & Francis.

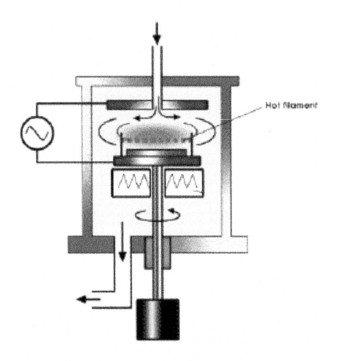

FIGURE 7.8

Schematic drawing of a CVD apparatus.

Reproduced from Nessim, G.D. (2010). Properties, synthesis, and growth mechanisms of carbon nanotubes with special focus on thermal chemical vapor deposition. Nanoscale 2, 1306–1323, with permission of The Royal Society of Chemistry.

7.3.3 CVD METHOD

The CVD method is carried out in a two-step process: First a catalyst such as Ni, Fe, or Co is deposited on a substrate where nucleation of the catalyst is carried out via chemical etching using ammonia as an etchant. Then, a gaseous carbon source such as methane, carbon monoxide, or acetylene is introduced into the reaction chamber where the carbon molecule is decomposed into carbon atoms, which then recombine in the form of NTs. The gas decomposition is carried out by an energy source such as plasma or heated coil. The temperature range used for the synthesis of the NTs is between 650°C and 900°C (Hirlekar et al., 2009) (Fig. 7.8).

7.4 PURIFICATION OF NTs

NTs usually contain a large amount of impurities such as metal particles, fullerenes, nanocrystalline graphite, and amorphous carbon. The different steps in purifying NTs are described below (Daenen et al., 2003).

7.4.1 **AIR OXIDATION**

CNTs that are less pure need to go through purification processes before the attachment of drugs onto CNTs. Air oxidation is effective in reducing the amount of amorphous carbon and metal catalyst particles such as Ni and Y. The optimal oxidation condition is found to be at 673 K for 40 min (Daenen et al., 2003).

7.4.2 **ACID REFLUXING**

Refluxing NTs in strong acid such as HCl, HNO_3, and H2SO4 is found to be useful in reducing the amount of metal particles and amorphous carbon. HCl is the ideal refluxing acid (Daenen et al., 2003).

7.4.3 **SURFACTANT-AIDED SONICATION, FILTRATION, AND ANNEALING**

Surfactant-aided sonication is usually carried out after the acid refluxing method, as the tubes become entangled together, trapping most of the impurities, such as carbon particles and catalyst particles after refluxing. Sodium dodecyl benzene sulfonate aided sonication with either ethanol or methanol as organic solvent is preferred because it takes the longest time for CNTs to settle down, indicating an even suspension state is achieved. The sample is then filtered with an ultra-filtration unit and annealed at 1270 K in N_2 for 4 h (Daenen et al., 2003). Annealing is found to be effective in optimizing the CNT structures as well as the surfactant-aided sonication in untangling the NTs and freeing the particulate impurities embedded in the entanglement.

Table 7.3 shows a summary of the three most common techniques used in making CNTs.

Table 7.3 Summary of the Major Production Methods of NTs

Method	Arc Discharge Method	Chemical Vapor Deposition	Laser Ablation
Who	Ebbesen and Ajayan, NEC, Japan in 1992	Endo, Shinshu University, Nagano, Japan	Smalley, Rice in 1995
How	Connect two graphite rods to a power supply, place them a few millimeters apart, and throw the switch. At 100 A, carbon vaporizes and forms hot plasma	Place substrate in oven, heat up to between 650°C and 900°C, and slowly add a carbon-bearing gas such as methane. As gas decomposes, it frees up carbon atoms, which recombine in the form of NTs	Blast graphite with intense laser pulses; use the laser pulses rather than electricity to generate carbon gas from which the NTs form; try various conditions until hit on one that produces prodigious amounts of SWNTs
Typical yield	30–90%	20–99%	Up to 70%
SWNT	Short tubes with diameters of 0.6–1.4 nm	Long tubes with diameters ranging from 0.6 to 4 nm	Long bundles of tubes of 5–20 μm, with individual diameter from 1 to 2 nm
MWNT	Short tubes with inner diameter of 1–3 nm and outer diameter of approximately 10 nm	Long tubes with diameter ranging from 10 to 240 nm	Not much interest in this technique, as it is too expensive, but MWNT synthesis is possible

(Continued)

Table 7.3 Summary of the Major Production Methods of NTs (Continued)

Method	Arc Discharge Method	Chemical Vapor Deposition	Laser Ablation
Pros	Can easily produce SWNT and MWNT. SWNTs have few structural defects; MWNTs without catalyst, not too expensive	Easiest to scale up to industrial production; long length, simple process, SWNT diameter controllable, quite pure	Primarily SWNTs, with good diameter control and few defects. The reaction product is quite pure
Cons	Tubes tend to be short with random sizes and directions; often needs a lot of purification	NTs are usually MWNTs and often riddled with defects	Costly technique, because it requires expensive lasers and high power requirement, but it is improving

From Daenen, M., et al. (2003). The wondrous world of carbon nanotubes. Eindhoven University of Technology.

7.5 APPLICATIONS OF CNTs

CNTs are one of the most important materials under investigation for nanotechnology applications. Their unique properties, ranging from ultrahigh strength through unusual electronic behavior and high thermal conductivity to an ability to store nanoparticles inside the tubes themselves, have suggested potential applications in many different fields of scientific and engineering endeavors. Key areas of existing and potential CNT applications include electronics, sensors, structural materials, fillers, and storage materials. The most highly developed commercial application for this material is the use of MWNT as a filler material in plastic composites and paints, sometimes as an improved substitute for carbon black. This market has been identified as having a multibillion dollar value. The extent of applications for CNT will depend on improvements in synthesis methods. Reducing costs is the most crucial factor, but the ability to regularly produce tubes with high lengths, tubes with specific chirality, individual tubes or ropes that are highly commensurate, and tubes with specific electronic properties will all be necessary for different commercial applications. Currently it appears that field emission displays are likely to be the first widespread use of CNT beyond the current application of MWNT as a filler material. Other applications will depend both on further research on the applications and substantial improvements in CNT synthesis methods (Baugham et al., 2002; Makar & Beaudoin, 2003).

7.5.1 EXISTING APPLICATIONS

7.5.1.1 CNT composites

Outstanding mechanical properties of CNT have sparked great interest in their use as structural materials. CNTs can be used as reinforcing fibers due to their strength and ultrahigh aspect ratios. Emphasis has been put on polymer composites with a number of commercial CNT reinforced resins already available on the market. Achieving proper dispersion and suitable CNT-matrix bonding are the two major ongoing challenges in this area of research. Dispersion is much more complex than simply mixing a powder of NTs into the liquid matrix material because CNTs tend to adhere together after purification. Functionalization of the tubes and the use of surfactants in combination with sonication are some of the methods developed to control dispersion. Typical CNT polymer composites experience fiber pullout under low loads and do not achieve high strengths and so research is ongoing in this area (Baugham et al., 2002).

Advances have also been made in development of CNT-metal and CNT-ceramic composites. Complications arise due to the higher temperatures needed to sinter the matrix materials. However, CNT-aluminum composites do not seem to experience the carbide formation seen in carbon fiber composites. A strengthened alumina ceramic is made by adding alumina/ethanol slurry to CNT dispersed in ethanol. The resulting powder is then sieved and ball milled to produce evenly distributed CNT followed by spark sintering. The initial stages produce evenly distributed CNT, while the sintering method ensures a fully dense material while maintaining nanometric grain sizes in the alumina and avoiding damaging the CNT (Kuzumaki et al., 2002; Ma et al., 1998).

7.5.1.2 Field emission displays

Research on electronic devices has focused primarily on using SWNTs and MWNTs as field emission electron sources for flat panel displays, lamps, gas discharge tubes providing surge protection, and X-ray and microwave generators. As a result of the small radius of the nanofiber tip and the length of the nanofiber, a potential applied between a CNT-coated surface and an anode produces high local fields that cause electrons to tunnel from the NT tip into the vacuum. Electric fields direct the field-emitted electrons toward the anode, where a phosphor produces light for the flat panel display application as shown in Fig. 7.9. NT tip electron emission arises from discrete energy states, rather than continuous electronic bands emitted from ordinary bulk materials. In addition, the emission behavior depends

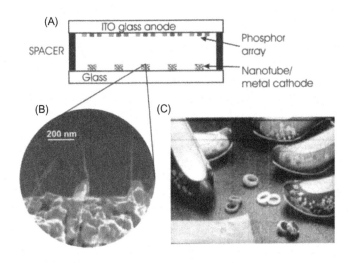

FIGURE 7.9

(A) Schematic illustration of a flat panel display based on CNTs. ITO, indium tin oxide. (B) Scanning electron microscopy (SEM) image of an electron emitter for a display, showing well-separated SWNT bundles protruding from the supporting metal base. (C) Photograph of a 5-in. (13-cm) NT field emission display made by Samsung.

From Baugham et al., 2002. Carbon nanotubes—the route toward applications. doi:10.1126/science.1060928.

critically on the NT tip structure: enhanced emission results from opening NT tips. Compared to tungsten and molybdenum tip arrays, NT field-emitting surfaces are easier to manufacture and require less vacuum for operation. NTs provide stable emission, long lifetimes, and low emission threshold potentials (Baugham et al., 2002).

Flat panel displays are one of the most promising applications being developed by industry. However, their commercialization is currently hindered by technical complexity of the device requiring concurrent advances in electronic addressing circuitry, the development of low-voltage phosphors, methods for maintaining the required vacuum, spacers withstanding the high electric fields, and the elimination of faulty pixels. The advantages of NTs over liquid crystal displays are low power consumption, high brightness, a wide viewing angle, a fast response rate, and a wide operating temperature range. Samsung has produced several prototypes including a 9-in. (23-cm) red-blue-green color display that can reproduce moving images (Baugham et al., 2002).

NT-based gas discharge tubes may soon find commercial use for protecting telecommunication networks against power surges. NT-containing cathodes separated from an anode by a millimeter-wide argon-filled gap provided a 4- to 20-fold improvement in breakdown reliability and a 30% decrease in breakdown voltage, as compared to commercial devices. When the phosphorescent screen at the anode in a field emission device is replaced by a metal target and the accelerating voltage is increased, X-rays are emitted instead of light. The resulting X-ray source has provided improved quality images of biological samples. The compact geometry of NT-based X-ray tubes suggests their possible use in X-ray source arrays for medical imaging, possibly even for X-ray endoscopes for medical exploration and microwave generation (Baugham et al., 2002).

7.5.1.3 Nanoelectronic devices

Radically different device materials, architectures, and assembly processes are developed in order for electronic circuits to continue to shrink by orders of magnitude and provide corresponding increases in computational power. Dramatic recent advances have fueled speculation that NTs will be useful for downsizing circuit dimensions. Current-induced electromigration causes conventional metal wire interconnects to fail when the wire diameter becomes too small. The covalently bonded structure of CNTs militates against similar breakdown of NT wires, and because of ballistic transport, the intrinsic resistance of the NT essentially vanishes (Baugham et al., 2002).

In nanotube field-effect transistors (NT-FETs), gating is achieved by applying a voltage to a submerged gate beneath an SWNT (as shown in Fig. 7.10A and B) which is contacted at opposite NT ends by metal source and drain leads. The transistors are fabricated by lithographically applying electrodes to NTs that are either randomly distributed on a silicon substrate or positioned on the substrate with an atomic force microscope. It is possible to selectively peel outer layers from an MWNT until an NT cylinder with the desired electronic properties is obtained as shown in Fig. 7.10C (Baugham et al., 2002).

Great advances have been made in research toward nanoscopic NT-FETs, which aims to replace the source-drain channel structure with an NT. The approach is to construct entire electronic circuits from interconnected NTs. A diode has to be produced by grafting a metallic NT to a semiconducting NT as shown in Fig. 7.11. The electronic properties depend on helicity, which can be tailored on both sides to produce a kinked structure. A revolutionary advance for nanoelectronics would be the development of rational synthesis routes to multiply branched and interconnected low-defect NTs with targeted helicity (Baugham et al., 2002).

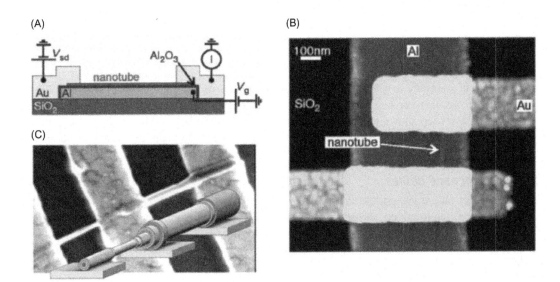

FIGURE 7.10

Nanoelectronic devices: (A) Schematic diagram for a carbon NT-FET. The semiconducting NT, which is on top of an insulating aluminum oxide layer, is connected at both ends to gold electrodes. The NT is switched by applying a potential to the aluminum gate under the NT and aluminum oxide. V_{sd}, source-drain voltage; V_g, gate voltage. (B) Scanning tunneling microscope (STM) picture of an SWNT field-effect transistor made using the design of (A). The aluminum strip is overcoated with aluminum oxide. (C) Image and overlaying schematic representation for the effect of electrical pulses in removing successive layers of an MWNT, so that layers having desired transport properties for devices can be revealed.

From Baugham et al., 2002. Carbon nanotubes—the route toward applications. doi:10.1126/science.1060928.

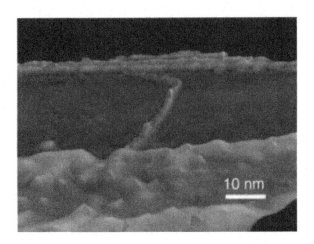

FIGURE 7.11

Nanoelectric device: STM image of an NT having regions of different helicity on opposite sides of a kink, which functions as a diode; one side of the kink is metallic and the opposite side is semiconducting.

From Baugham et al., 2002. Carbon nanotubes—the route toward applications. doi:10.1126/science.1060928.

7.5.1.4 Hydrogen storage

NTs have been in a lengthy development process as potentially useful material for hydrogen storage for fuel cells. However, there has been a lot of controversy regarding claims of high hydrogen storage levels that have been shown to be incorrect which await confirmation. More recent research has suggested that CNTs are unlikely to be effective storage devices (Baugham et al., 2002; Makar and Beaudoin, 2003).

7.5.1.5 Sensors and probes

Because NT electronic transport and thermopower (voltages between junctions caused by interjunction temperature differences) are very sensitive to substances that affect the amount of injected charge, chemical sensor of nonmetallic NTs hold great potential for state-of-the art sensor applications. The minute size of the NT sensing element and the correspondingly small amount of material required for a response are the main advantages of CNTs. However, major challenges remain in making devices that differentiate between absorbed species in complex mixtures and provide rapid forward and reverse responses. CNT scanning probe tips for atomic probe microscopes are now commercially available from Seiko Instruments. The mechanical robustness of the NTs and the low buckling force dramatically increase probe life and minimize sample damage during repeated hard crashes into substrates. The cylindrical shape and small tube diameter enable imaging in narrow, deep crevices and improve resolution in comparison to conventional nanoprobes, especially for high sample feature heights. Mapping of chemical and biological functions is enabled by covalently modifying the NT tips, such as by adding biologically responsive ligands. Nanoscopic tweezers that may be used as nanoprobes for assembly are driven by the electrostatic interaction between two NTs on a probe tip (Baugham et al., 2002).

7.5.2 POTENTIAL AND UNDER DEVELOPMENT APPLICATIONS

7.5.2.1 Body armor

The University of Cambridge has developed a new type of high strength and lightweight fiber from CNTs which has promising applications for body armor to be used for the military or the police. These fibers can be woven as a cloth or can be incorporated into composite materials to produce super-strong products. In addition to the very high strength of this material, the CNT fibers are able to absorb and distribute the impact (Gordeyev, 2010).

These fibers synthesized with a hydrocarbon material such as ethanol alongside an iron-based catalyst are sent to a furnace. The hydrocarbon breaks down into its hydrogen and carbon atoms and then the free carbon binds to particles of the catalyst in the form of long NTs. Production of the NTs in the reaction chamber results in formation of a smoke-like super-elastic material. In the next step, a rod is inserted into the furnace to grab one of the formed NTs' socks to stretch them into a filament that can be wound up continuously on a reel. This act increases the intermolecular forces which hold the tubes together and increases the strength of the overall material (Gordeyev, 2010).

This method is currently under investigations to be upgraded to industrial scale. UK's Ministry of Defence and the US army are looking into this project to use these fibers for new applications such as hi-tech smart cloth or bomb-proof refuse bins.

7.5.2.2 Space elevator

In a joint project between the University of Cambridge and NASA, the researchers have developed a method to combine multiple separate NTs together to form light, flexible, and long strands. These long strands could be used to bring space elevators closer to reality.

In theory, in order to build a space elevator, a cable would extend 22,000 miles above the Earth surface at the Geo-stationary orbit of the Earth. At this distance the elevator station would remain stationary like a satellite. The cable is then extended for another 40,000 miles into space to a weighted structure for stability. An elevator car would be attached to the NT cable and powered into space along the track (Edwards, 2000).

NASA's shuttle fleet retired in 2010 and it cannot be replaced until 2014 due to insufficient funding. CNT cables are a promising and cost-effective method to provide transportation to the international Space Station. However, to fulfill NASA's need for 144,000 miles of NTs, commercial scale production of NTs will be required (NASA Science, 2000).

7.5.2.3 Artificial muscle

Researchers at Florida State University have developed a novel method to produce CNT aerogel sheets via a solid-state process. These solid-state fabricated sheets, which are the sole components of new artificial muscle, provide giant elongations and elongation rates of 220% and 3.7×10^4% per second, respectively, at operating temperatures from 80 to 1900 K (Aliev, 2009).

The researchers have grown forests of 11-nm diameter CNTs and then pulled the tubes into ribbons composed of oriented bundles. Because of special alignment of the NTs arrays they can be pulled into sheets at speeds of up to 2 m/s. These sheets have very low density and have a very high specific strength in stretch direction. However, in other directions, they are very fragile.

Electrostatic repulsion between NTs makes the sheets able to expand to up to three times their original size when a positive voltage is applied and to shrink back down to their original size when the voltage is shut off. Fig. 7.12 shows an artificial muscle expansion at room temperature (Fig. 7.12B) and at 1500 K (Fig. 7.12C) while applying voltage of 5 kV.

Having the same cross-sectional area, these artificial muscles made of aerogel sheets can generate 30 times the force of a natural muscle. In addition, these artificial muscles can elongate 10 times more than natural muscle at 1000 higher rate (Aliev, 2009).

7.5.2.4 Light bulb filament

Researchers at Louisiana State University have developed light bulb filaments by immersing CNTs in alcohol and then assembling them into long filaments under surface tension when the alcohol is evaporated. The NT filaments were connected to the electrodes and sealed in a glass bulb under vacuum.

NT filaments have shown to have lower threshold voltage for incandescent light emission than the tungsten filament. For instance, a DWNT filament with a resistance of about 9 Ω begins to emit incandescent light at 3 V, an SWNT filament (18.2 Ω) begins to emit at 5 V, while tungsten filament (3 Ω) begins to emit at 6 V. Fig. 7.13 shows the comparison of irradiance intensity of the DWNT and the tungsten filaments as a function of voltage. It is observed that the NT bulbs have lower threshold voltage than the tungsten bulb. The irradiance intensity of the NT bulb increases quickly with increase in voltage. In addition, the irradiance intensity of the NT filament is much stronger than that of the

(A)   (B)   (C)

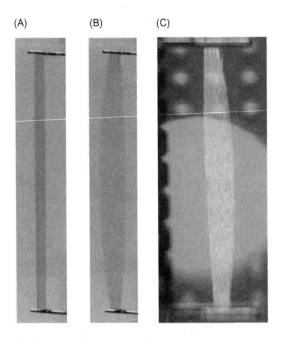

FIGURE 7.12

(A) An artificial muscle strip with no voltage applied. (B) The above artificial muscle strip with 5 kV applied.
(C) An artificial muscle strip actuated at 1500 K using 5 kV applied voltage.

From Aliev, A. E. (2009). Giant-stroke, superelastic carbon nanotube aerogel muscles. Materials Science.

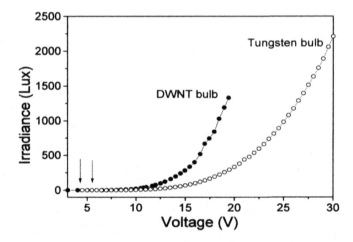

FIGURE 7.13

Irradiance of NT filaments as a function of voltage. The DWNT filament (9 Ω) shows a low onset voltage
(marked arrow) for the light emission and emits stronger light than tungsten filament (3 Ω) at the same voltage.

From Wei, J. (2004). Carbon nanotube filaments in household light bulbs. Applied Physics Letters.

tungsten, indicating that the NT filaments can emit more visible light than tungsten at the same applied voltage (Wei, 2004).

7.5.2.5 Solar cells

Based on a research at MIT, CNTs could be used to form antennas that capture and concentrate solar energy 100 times more than a photovoltaic (PV) cells. In fact, for the first time researchers have been able to construct NT fibers in which they can control the properties of different layers. This has been made possible by recent advances in separating NTs with different properties. This founding promises the possibility to make much smaller and more powerful solar arrays (Han, 2010).

These antennas including a fibrous rope are made of two layers of NTs with different electrical bandgaps. The inner layer of the antenna contains NTs with a lower bandgap compared to the one in the outer layer. Since the excitons flow from high to low energy, the excitons in the outer layer flow to the inner layer. When the material is struck by light, all of the excitons flow and concentrate at the center of the fiber. In fact, by constructing the antenna around the core of a semiconductor material, the antenna would concentrate photons before the PV cell converts them to an electrical current (Han, 2010).

7.5.2.6 Loud speakers

Researchers from the Nanotechnology Research Centre in Beijing developed loudspeakers from sheets of parallel carbon tubes, each about 10 nm across. These NT sheets can create sounds as loud as commercial speaker when an electric current alternating at an audio frequency is applied to them. The NT loudspeakers can be stretched up to twice their original size with little change to the intensity of the sound (Xiao, 2008).

Because of transparency and high flexibility of the NT sheets, they can be placed nearly anywhere. Researchers have already attached a transparent film to the screen of an iPod to play music from the device.

Applying an electric current to the NT films results in the generation of heat, which causes the expansion of the surrounding air and sound waves are created. This process is very similar to how lightning generates thunder with an exception that thunder is not a controlled discharge while the electrical discharge in the NT films can be controlled (Xiao, 2008).

7.5.2.7 Displays

Samsung demonstrated the world's first CNT-based color active matrix electrophoretic display (EPD) made e-paper in 2008. The new color e-paper device has a 14.3 in. format display.

In order to make these new displays it is required to create conductive NT films analogous to ITO technology which is a transparent semiconducting material used as an electrode on flat panel displays. In addition, it is required to have evenness over large areas in films and to have compatibility with different display technologies and fabrication processes.

The EPDs have several advantages over the old flat panel displays such as low power consumption and bright light readability. In addition, the image on the display is retained without the need to constantly refresh. The EPDs can be produced on thin and flexible substrate which makes them ideal for handheld and mobile applications. In near future, these films will be produced for various types of touch screen devices and can be applied for portable and flexible computers, cell phones, personal digital assistants, and many other applications. Furthermore, this technology has the potential to be used in plastic solar cells and organic LED lighting (Henry, 2008).

7.5.2.8 Nanoradio

Researchers from University of California developed a radio device using CNTs which can perform all four functionalities of a regular radio device. This small and simple structured device is an antenna, amplifier, demodulator, and tuner at the same time.

The researchers accumulated MWCNTs on a silicon electrode and connected that using two wires to a counter electrode at a micrometer away. To create a small field emission current between the NT tip and the counter electrode a DC battery was attached to the apparatus. The researchers placed the apparatus under a high-resolution transmission electron microscope to observe the function of this radio during the course of a radio transmission (Jensen, 2007).

Due to smaller size, less complex structure, lower power requirement, and biocompatibility, the radio potentially can be placed inside body for various diagnostic, therapeutic, monitoring, and sensory functions. In addition, the NT may be altered by contact with particles at atomic scale that change the resonance frequency of the NT. This change can be used to create high-sensitive mass spectrometers which are able to detect the mass of less than a single hydrogen atom (Jensen, 2007).

7.5.2.9 Bucky paper

Bucky papers are thin films made of NTs aggregates. These films are exceptionally lightweight ($21.5\,g/m^3$), highly flexible with nanoscale porous structures. Bucky papers provide the ability to effectively transfer the properties of NTs into composites. Direct mixing of NTs into polymer matrices causes the NTs to group together and not disperse in the composite (Genuth, 2006).

Bucky papers are synthesized by suspension of CNTs in an aqueous solution using nonionic surfactants such as Trition X-100. The suspension is filtered using membrane to yield uniform films of pure NTs. These films or Bucky papers have strength higher than diamond at a fraction of the weight.

The current production level of Bucky paper is at lab scale only, however, the researchers are planning to develop a prototype for continuous production at large scale. By scaling up the production level and decreasing the cost, the Bucky papers are prospected to be initially in military applications. The Bucky papers can be used to build aircrafts with electromagnetic interference shielding and lightning strike protection. In addition, Bucky papers can be used in automotive industry to build stronger cars while having less weight and more fuel efficiency (Genuth, 2006).

7.6   CURRENT RESEARCH AND FUTURE PERSPECTIVES

NTs are an important part of nanotechnology with a variety of existing and potential applications. Their properties are of much value and are heavily researched. As with other areas of research within this field, the use of NTs in biomedicine may prove to be the most rewarding application. NTs are often the topic of many studies done on cancer treatment.

For example, Kalbacova (2008) has demonstrated the use of self-organized TiO_2 NT layers as photocatalytic killing subjects of HeLa G cancer cells. They used TiO_2 NT layers with a diameter of 50 nm and thickness of 800 nm and another set with a diameter of 100 nm and a thickness of 1.3 μm, which were grown by anodization of titanium. After incubating the HeLa G cancer cells and UV exposure (Fig. 7.14), the changes in the cell morphology and viability were measured and it was found that the vitality of the cancer cells cultured on these NT layers was significantly affected. It was observed that the shape and size of the cells were reduced and a significant amount of the dead cells was found (Fig. 7.15). These results demonstrate that self-organized TiO_2 NT layers can be used for photo-induced cancer cell killing (Kalbacova, 2008).

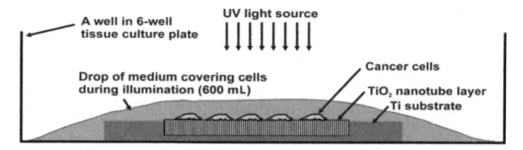

FIGURE 7.14

Scheme of the setup used during the experiments.

From Kalbacova (2008). TiO₂ nanotubes: photocatalyst for cancer cell killing. Physica Status Solidi, 194–196.

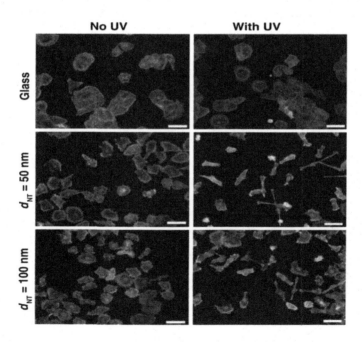

FIGURE 7.15

Fluorescence microscopy images of the morphology of HeLa G cells cultured on TiO_2 NT layers (diameter of NTs, d_{NT}=50 and 100 nm) and glass with and without UV irradiation. Actin filaments are stained green (light gray in print version) and nuclei blue (dark gray in print version). The scale bars are 50 μm. As it can be seen, cells on TiO_2 NT layers showed significant size reduction after UV irradiation while cells cultured on the glass control did not exhibit dramatic changes even after exposure to UV light.

From Kalbacova (2008). TiO₂ nanotubes: photocatalyst for cancer cell killing. Physica Status Solidi, 194–196.

Another interesting study for the treatment of cancer was done by Xiao in 2009. They describe a method for creating dual-purpose HER2 IgY–SWNT complex conjugated from carboxylated SWNTs with anti-HER2 chicken IgY antibody for both detection and selective destruction of cancer cells in an in vitro model consisting of HER2-positive SK-BR-3 cells and HER2-negative MCF-7 cells. HER2 IgY–SWNT complex targets HER2-expressing SK-BR-3 cells but not receptor-negative MCF-7 cells. They used near-infrared (NIR) laser light at a wavelength of 785 nm which is reflected intensely off the NTs and can be detected by Raman spectroscopy. Researchers then increased the laser wavelength to 808 nm. This is absorbed by the NTs and incinerates the NTs and the HER2 tumor cells attached to them. Fig. 7.16 shows the result of NIR irradiation with an 808 nm laser at 5 W/cm^2 for 2 min. The alphabetical letters represent the following: (A) untreated SK-BR-3 cells, (B) SK-BR-3 cells treated with SWNT alone, (C) SK-BR-3 cells treated with anti-HER2 IgY antibody alone, (D) SK-BR-3 cells treated with the HER2 IgY–SWNT complex, (E) untreated MCF-7 cells, (F) MCF-7 cells treated with the HER2 IgY–SWNT complex. In these figures, cells with green (light gray in print version) fluorescence were considered alive, whereas those with red (dark gray in print version) fluorescence were dead. Bar graphs on figure (G) show the percentage of live cells in each sample of SK-BR-3 cells following NIR irradiation (Xiao, 2009).

The above results demonstrate the high transparency of biosystems to NIR light with wavelength of 808 nm at 5 W/cm^2 for 2 min. SK-BR-3 cells treated with the HER2 IgY–SWNT complex showed

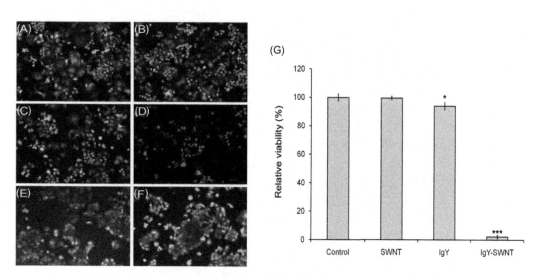

FIGURE 7.16

Cell viability after treatment with the HER2 IgY-SWNT complex followed by NIR irradiation with an 808 nm laser at 5 W/cm^2 for 2 minutes. (A) Untreated SK-BR-3 cells, (B) SK-BR-3 cells treated with SWNT alone, (C) SK-BR-3 cells treated with anti-HER2 IgY antibody alone, (D) SK-BR-3 cells treated with the HER2 IgYSWNT complex, (E) Untreated MCF-7 cells, (F) MCF-7 cells treated with the HER2 IgY-SWNT complex, and (G) Bar graphs show the percentage of live cells in each sample.

From Xiao, Y. (2009). Anti-HER2 IgY antibody-functionalized single-walled carbon nanotubes for detection and selective destruction of breast cancer cells. Chemical Science and Technology Laboratory, National Institute of Standards and Technology (NIST).

extensive cell death after heating with NIR irradiation (Fig. 7.16D and G). In stark contrast, negligible cell death was observed with SK-BR-3 cells treated with SWNTs alone (Fig. 7.16B and G) or untreated (Fig. 7.16A and G), and in MCF-7 cells treated with the HER2 IgY–SWNT complex (Fig. 7.16F) (Xiao, 2009.).

Different methods for drug delivery have been discussed numerous times throughout this book. NTs can be used in certain processes of drug delivery to help cancer patients. Zhuang Liu (2008) examined the ability of chemical functionalized SWCNTs for drug delivery to suppress tumor cells in female mice bearing inoculated 4T1 tumors. Water soluble SWNT–paclitaxel (PTX) conjugate was obtained through conjugating PTX to branched polyethylene glycol (PEG) chain on SWNT via a cleavable ester bond. The authors performed their experiment by treating the mice through injecting Taxol, PEG–PTX, DSEP–PTX, and SWNT–PTX. They observed that after 22 days, the SWNT–PTX resulted in significant difference in average fractional tumor volume compared to the other treatment methods. Furthermore, they tested the SWNT delivery of PTX into xenograft tumors in mice through treating the tumor cells by SWNT–PXT versus Taxol and Plain SWNT. SWNT–PTX showed high tumor suppression as only ~20% of proliferation active cells were noted, whereas Taxol and Plain SWNT showed no change in the number of tumor cells. Fig. 7.17 shows the change in the relative tumor volume (calculated as V/V_0, where V_0 is the tumor volume when the treatment was initiated). The same

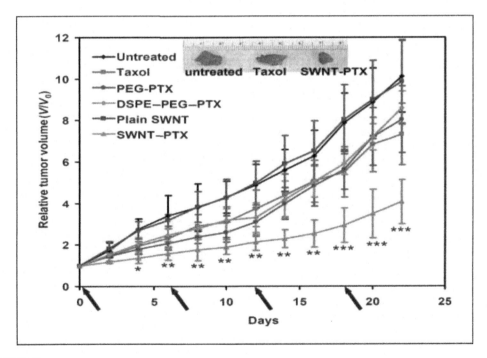

FIGURE 7.17

NT PTX delivery suppresses tumor growth of 4T1 breast cancer mice model.

From Zhuang Liu, K. C. (2008). Drug delivery with carbon nanotubes for in vivo cancer treatment. The Molecular Imaging Program at Stanford; Department of Chemistry, Stanford University.

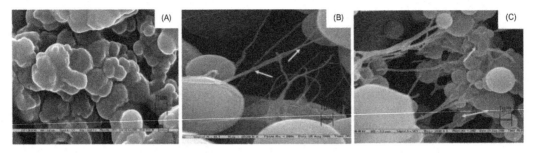

FIGURE 7.18

SEM micrographs of (A) SiO_2, (B) MWCNTs/SiO_2, and (C) Urs/MWCNTs/SiO_2 at 20,000 magnification.

From Ahuja, T. (2010). Potentiometric urea biosensor based on multi-walled carbon nanotubes (MWCNTs)/silica composite material. National Physical Laboratory (Council of Scientific & Industrial Research).

PTX dose (5 mg/kg) was injected (on days 0, 6, 12, and 18, marked by arrows) for Taxol, PEG–PTX, DSEP–PEG–PTX, and SWNT–PTX (Zhuang Liu, 2008).

PTX conjugated to PEGylated SWNTs exhibits high water solubility and maintains similar toxicity to cancer cells as Taxol in vitro. However, SWNT–PTX affords much longer blood circulation compared to Taxol and PEGylated paclitaxel (PEG-PTX), causing higher tumor uptake of the drug through the Enhanced Permeability and Retention (EPR) effect. Consequently, the SWNT–PTX is able to slow down tumor growth even through a low drug dose. The non-Cremophor composition of SWNT–PTX, rapid clearance of drugs from reticuloendothelial system (RES) organs, higher ratios of tumor-to-normal organ drug uptakes, and the fact that tumor suppression efficacy can be reached at low injected drug doses make CNT drug delivery a very promising nano-platform for future cancer therapeutics (Zhuang Liu, 2008).

NTs may also be used in the treatment of other health problems. Urea is the most important product of protein degradation, and its levels in the blood are often checked during clinical diagnostics for a variety of problems stemming from issues with renal health (Ahuja, 2010). Ahuja (2010) describes a method for fabricating a urea biosensor with "Urease (Urs) immobilized MWCNTs embedded in silica matrix deposited on the surface of indium tin oxide coated glass plate" to provide a biocompatible environment for entrapment of Urs enzyme. The Urs is covalently linked with the exposed –COOH groups of the functionalized multi-walled carbon nanotubes (F-MWCNTs), which are subsequently incorporated within the silica matrix. The Urs/MWCNTs/SiO_2/ITO composite modified electrode was characterized by Fourier transform infrared spectroscopy, thermal gravimetric analysis, and UV–visible spectroscopy. The physical morphology of the modified Urs/MWCNTs/SiO_2/ITO electrode was investigated by comparing SiO_2 and F-MWCNT/SiO_2 with and without Urs enzyme using SEM. Fig. 7.18A is the SEM micrograph of silica. Fig. 7.18B is the SEM micrograph of F-MWCNT/SiO_2. It shows fine nanotubular structures embedded within the circular particles of the silica matrix. Fig. 7.18C shows the SEM micrograph of enzyme immobilized Urs/MWCNTs/SiO_2 composite having aggregation of globular-shaped enzyme molecules surrounding the NTs in the silica matrix (Ahuja, 2010).

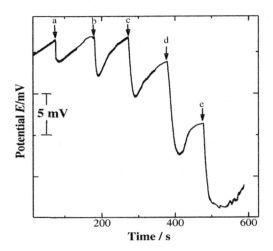

FIGURE 7.19

Chronopotentiometric response of biosensor Urs/MWCNTs/SiO$_2$/ITO to increasing urea concentrations.

From Ahuja, T. (2010). Potentiometric urea biosensor based on multi-walled carbon nanotubes (MWCNTs)/silica composite material.
National Physical Laboratory (Council of Scientific & Industrial Research).

The electrochemical performance of the modified electrode was tested by a potentiometric method after the addition of urea buffer solution which resulted in a quick response of the enzyme electrode to the urea. Fig. 7.19 shows the chronopotentiometric response of biosensor Urs/MWCNTs/SiO$_2$/ITO to increasing urea concentrations: (A) 3.1×10^{-5} M, (B) 1×10^{-4} M, (C) 3.1×10^{-4} M, (D) 1×10^{-3} M, and (E) 3.1×10^{-3} M. The arrow indicates the time at which the aliquot of urea was dispensed in the sample solution (Ahuja, 2010).

The synergistic effect of silica matrix, F-MWCNTs, and the biocompatibility of Urs/MWCNTs. SiO$_2$ allowed the biosensor to have a wide linear range urea detection of 2.18×10^{-5} to 1.07×10^{-3} M urea. The biosensor shows a short response time of 10–25 s and a high sensitivity of 23 mV/decade/cm^2 (Ahuja, 2010).

The science of biology, in general, is a field that often deals with small molecules that may be interacted with using nanotechnology. Wilner and Henning (2010) present a new method to synthesize nonseparable covalently linked DNA NTs. They used a circular DNA that includes at its opposite poles thiol and amine functionalities that act as the building block for the construction of the DNA NTs. The circular DNA is cross-linked with a bis-amide-modified nucleic acid to yield DNA nanowires, and these are subsequently cross-linked by a bis-thiolated nucleic acid to yield the DNA NTs. Alternatively, a circular DNA that includes four amine functionalities on its poles is cross-linked in one step by the bis-thiolated nucleic acid to yield the NTs. Fig. 7.20 shows a schematic formation of the resulting NTs which were found to be stable, stiff, nonseperable, and could withstand high temperatures. Therefore, they are physically and chemically sturdy enough to withstand the manipulations necessary for many technological applications (Wilner & Henning, 2010).

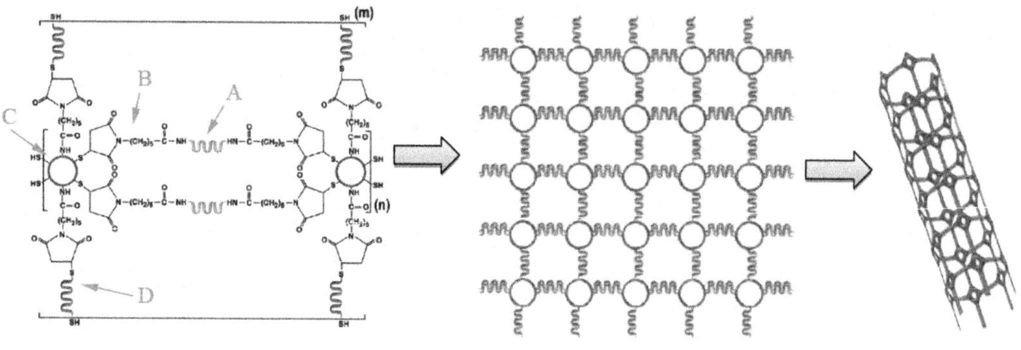

FIGURE 7.20

Synthesis of covalently linked DNA NTs through the orthogonal stepwise cross-linking of bis-thiolated/bis-aminated circular DNAs, where A is the nucleic acid functionalized at its 3′ and 5′ ends with amine functionalities; B is the bifunctional reagent *N*-[(ε-maleimidocaproyloxy) sulfosuccinimide ester)] (Sulfo-EMCS); C is the circular DNA which is modified at its north/south poles with amine functionalities, whereas the east/west poles of the circular DNA are modified each with vicinal dithiol functions; and D is a nucleic acid, modified at its 3′ and 5′ ends with thiol functionalities to yield the covalently cross-linked two-dimensional array of the circular DNAs. This array could fold into a cylindrical NT as shown in the figure.

From Wilner, O. I., Henning, A. (2010). Covalently linked DNA nanotubes. American Chemical Society, 1458–1465.

Biomineralization is a method used to form hybrid inorganic/organic nanomaterials, and is a field common to biology, medicine, chemistry, and materials science (Shchukin & Sukhorukov, 2005). Shchukin and Sukhorukov (2005) employed commercially available and cheap aluminosilicate NTs—halloysites—as new biomineralization nanoreactors for carrying out enzyme-catalyzed inorganic synthesis inside the hollow tubular lumen. The synthesis procedure is shown in Fig. 7.21. The urease was adsorbed on the inside of the tubes by attraction between negatively charged enzyme and positive Al_2O_3 surface. After washing with distilled water, the halloysite loaded with urease was immersed into a mixture of urea and $CaCl_2$ for 30 min. The urease catalyzed the decomposition of urea into ammonia and CO_3^{2-} ions, which immediately reacted with Ca^+ ions resulting in formation of $CaCO_3$ precipitate exclusively inside the tubes achieved by maintaining high concentration of diffusions Ca^+ ions. The tubes were characterized by TEM, XRD, and SEM. The TEM image of the tubes cross section in Fig. 7.22 shows uniform filling of the tubular lumen with $CaCO_3$. Crystallization forces have also lead to swelling of the halloysite shell and the confined reaction micro (nano-) volume had a strong influence on the morphology of $CaCO_3$ resulting in the formation of a more stable calcite phase not found under any synthetic conditions. The techniques developed in this study may be utilized for loading the interior of halloysite with different compounds or with bioactive material. Moreover, there is a great prospect of fabricating complex inorganic core/shell-type nanomaterials composed of two completely different substances (Shchukin & Sukhorukov, 2005).

In addition to biology and medicine, NTs are researched for their applications for electricity and electronics. For example, NTs may be used to replace certain components in lithium ion batteries. Song (2010) developed a nanostructured form of silicon, consisting of arrays of sealed, tubular geometries that are capable of accommodating large volume changes associated with lithiation in battery

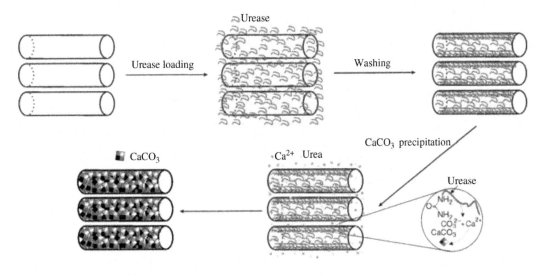

FIGURE 7.21

Schematic illustration of the urease-catalyzed synthesis of $CaCO_3$ inside halloysite NTs.

From Shchukin, D. G., Sukhorukov, G. B. (2005). Halloysite nanotubes as biomimetic nanoreactors. Small, 1, 510–513.

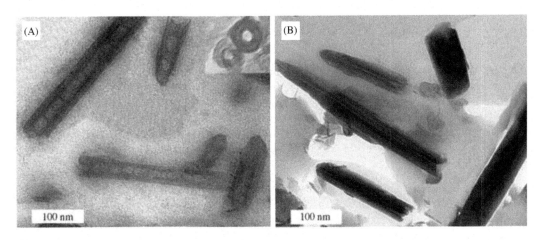

FIGURE 7.22

Longitudinal transmission electron microscopy images of ultra-microtome hallysite G NTs before (A) and after (B) $CaCO_3$ formation. The inset in (A) shows the equivalent perpendicular cross section.

From Shchukin, D. G., Sukhorukov, G. B. (2005). Halloysite nanotubes as biomimetic nanoreactors. Small, 1, 510–513.

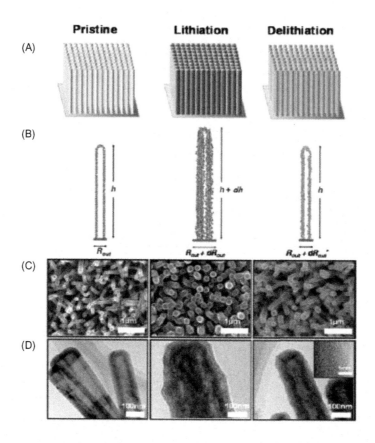

FIGURE 7.23

Schematic illustration and electron microscopy images of Si NTs before reaction and after full lithiation and delithiation. (A and B) By using an ordered sacrificial template, uniform free spaces inside the Si NTs and regular intertube spacings are possible. These free surfaces offer mechanics that facilitates expansion, as manifested in an anisotropic response. The free spaces radially decrease in size during lithiation and fully recover during delithiation. Microscopy shows the morphological changes during the first cycling; where dR_{out}^* is defined as the change in outer radius after cycling. (C) Top view SEM images and (D) TEM images.

From Song (2010). Arrays of sealed silicon nanotubes as anodes for lithium ion batteries.

applications. The silicon NTs developed were exhibiting high initial Coulombic efficiencies, which was found to be more than 85%, and stable capacity retention, which was found to be more than 80% after 50 cycles, due to an unusual, underlying mechanics that is dominated by free surfaces. This physics is manifested by a strongly anisotropic expansion in which 400% volumetric increases are accomplished with only relatively small changes in the axial dimension which was found to be less than 35%. These experimental results and associated theoretical mechanics models demonstrate the extent to which nanoscale engineering of electrode geometry can be used to advantage in the design of rechargeable batteries with highly reversible capacity and long-term cycle stability (Song, 2010) (Fig. 7.23).

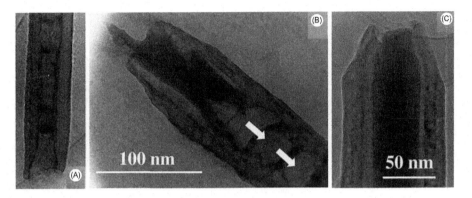

FIGURE 7.24

TEM images of MWNTs filled with nitrate salt solution (A) as is and after heating in air at (B) 50°C and (C) 100°C.

From Keller, N., Pham-Huu, C., Estournès, C., Grenèche, J.-M., Ehret, G., Ledoux, M. J., (2004). Carbon nanotubes as a template for mild synthesis of magnetic CoFe₂O₄ nanowires. Carbon, 42, 1395–1399.

NTs may also be used for data storage. Keller et al. (2004) used MWCNTs as nanoreactors for synthesis of $CoFe_2O_4$ nanowires encapsulated inside the NTs. MWNTs with mean outer and inner diameters of 100 and 60 nm, respectively, and up to several micrometers long were submerged into filling solution containing cobalt and iron nitrates with a molar ration of 1:2 and at a metal loading of 30 wt% relative to MWNTs. The MWNTs were completely filled by the nitrate solution due to capillary forces (Fig. 7.24A). After drying at 50°C formation of solid caps on both sides of the tube was observed (Fig. 7.24B) which lead the researchers to consider the inside of each NT like a closed reactor. The solid caps also hinder diffusion of oxygen, which competes for cobalt in oxidation reactions, inside of the tubes. Further calcinating the tubes at 100°C lead to an increase in pressure inside the NTs, which facilitated formation of $CoFe_2O_4$ nanowires along the entire length of the tubes with dendritic crystal structure (Fig. 7.24C). Further increasing temperature up to 600°C led to the modification of the nanowire microstructure into a well-crystallized nanowire composed of several single crystals along the tube axis (Fig. 7.25). The magnetic nanowires separated by nonmagnetic carbon walls could be used for high-density data storage without the drawback of particle agglomeration and magnetic losses due to dipolar relaxation (Keller et al., 2004).

NTs, as with other areas of nanotechnology, are researched heavily for their potential in finding new sources of energy or greener solutions. Hydrogen is expected to become a good source because it is clean burning. Eswaramoorthi, Sundaramurthy, and Dalai (2006) studied the feasibility of using CNTs as support to Cu-Zn catalysts for hydrogen production from partial oxidation of methanol (POM). CNTs were first functionalized by refluxing with nitric acid and sulfuric acid to enhance catalyst absorption. The catalyst was supported by mixing aqueous solution of nitrates of copper and zinc with CNTs and stirring for 2 h at 60–70°C. POM over Cu-Zn/CNTs catalysts was carried out in a fixed-bed reactor at atmospheric pressure between 220°C and 280°C. The metal catalyst was dispersed on the surface of CNTs in the form of spherical particles as shown by TEM images of Cu-Zn/CNTs in Fig. 7.26B–D. Metal loading had an effect on metal particle size and dispersion. The average particle

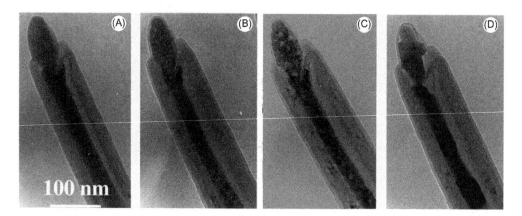

FIGURE 7.25

In situ TEM images recorded as a function of the in situ heating temperature on the sample first calcined in air at 100°C: (A) as is, (B) 450°C, (C) 500°C, and (D) 600°C showing densification of $CoFe_3O_4$ crystal structure with temperature.

From Keller, N., Pham-Huu, C., Estournès, C., Grenèche, J.-M., Ehret, G., Ledoux, M. J., et al. (2004). Carbon nanotubes as a template for mild synthesis of magnetic $CoFe_2O_4$ nanowires. Carbon, 42, 1395–1399.

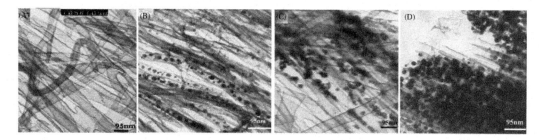

FIGURE 7.26

TEM images of (A) purified CNTs, (B) 5Cu-3Zn/CNTs, (C) 7Cu-5Zn/CNTs, and (D) 12Cu-9Zn/CNTs.

From Eswaramoorthi, I., Sundaramurthy, V., & Dalai, A. K. (2006). Partial oxidation of methanol for hydrogen production over carbon nanotubes supported Cu-Zn catalysts. Applied Catalysis A: General, 313, 22–34.

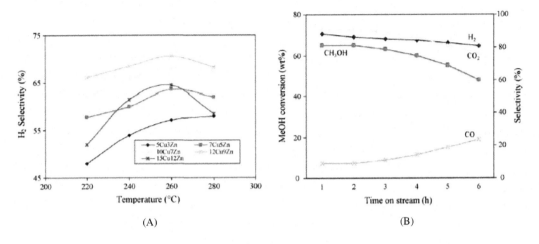

FIGURE 7.27

Effect of (A) temperature on hydrogen selectivity over different catalysts; (B) time on stream on methanol conversion, H_2, CO_2, and CO selectivity over 12 wt% Cu-9 wt%Zn/CNTs catalyst at 260°C.

From Eswaramoorthi, I., Sundaramurthy, V., & Dalai, A. K. (2006). Partial oxidation of methanol for hydrogen production over carbon nanotubes supported Cu-Zn catalysts. Applied Catalysis A: General, 313, 22–34.

size for 5Cu-3Zn/CNTs catalyst was 5.6 nm and that of 10Cu-7Zn/CNTs was 9.8 nm. A threshold metal loading was found to be 12Cu-9Zn/CNTs beyond which catalyst dispersion was reduced due to particle agglomeration. The results of POM are shown in Fig. 7.27. The optimum reaction temperature was found to be at 260°C. Sample 12 wt% Cu-9 wt% Zn/CNTs displayed most favorable catalytic properties due to the improved metal dispersion, narrow particle size distribution, and almost complete reduction of Cu particles and the catalyst activity dropped over time due to oxidation of Cu^0 to CuO (Eswaramoorthi et al., 2006).

NANOSHELLS*

8.1 INTRODUCTION

8.1.1 NANOTECHNOLOGY

Nanotechnology refers to the development and application of structures, materials, devices, and systems at the nanoscale, with sizes ranging between 1 and 100 nm. It involves the manipulation and creation of material structures in the atomic, molecular, and supramolecular fields. At the nanoscale, the characteristics of matter can be significantly different, particularly under the range of 10–20 nm, because of phenomena such as dominance of quantum effects, confinement effects, molecular recognition, and an increase in relative surface area. The downsized material structures of chemical elements have different mechanical, optical, magnetic, and electronic properties, and thus have surprising and unpredicted effects. Nanodevices exist in an area where the properties of matter are governed by a complex combination of classical physics and quantum mechanics. At the nanometer scale, manufacturing capabilities such as self-assembly, templating, stamping, ultra-precision engineering, and fragmentation are broad and can lead to numerous efficient outcomes. Nanotechnology requires the integration of many scientific, engineering, and technical disciplines and competences. Applications of nanotechnology have the potential to affect a wide variety of sectors, such as communication, health, labor, mobility, housing, relaxation, energy, and food. They will thus be accompanied by changes in the social, economic, ethical, and ecological spheres. Nanotechnology has the potential to become one of the defining technologies of the 21st century. Based on the ability to measure, manipulate, and organize material on the nanoscale, it is set to have significant implications (Pradeep, 2009). Breakthroughs in nanotechnology include order of magnitude increases in computer efficiency, advanced pharmaceuticals, biocompatible materials, nerve and tissue repair, surface coatings, improved catalysts, sensors, telecommunications, and pollution control.

8.1.2 NANOSHELLS

Nanoshells are important for cancer and spectroscopic applications. The method of synthesis of nanoshells is quite simple and includes one-step and two-step approaches. In the field of molecular encapsulation silica nanoshells are mainly used, while metal nanoshells are used for cancer therapy. The principal tools for their characterization are absorption spectroscopy, fluorescence spectroscopy,

---

*By Yaser Dahman, Nabeel Ashfaq, Ahilan Ganesalingam, and Ganeshakumar kobalasingham.

Nanotechnology and Functional Materials for Engineers. DOI: http://dx.doi.org/10.1016/B978-0-323-51256-5.00008-3

and transmission electron microscopy (TEM). Nanoparticles are stabilized by different types of ligands, such as organic molecules, polymers, and surfactants. Nanoparticles that are called core–shell particles are those whose surface is passivated by a shell with its own distinct properties, other than the core. This shell can be made of metals and oxides depending upon the use. This type of coating not only stabilizes colloidal dispersions, but also allows modification and tailoring of particle properties, such as optical, magnetic, and catalytic. Oxide-protected nanoparticles are found to be more stable in extreme conditions. During exposure to intense lasers, for example, these particles are more resistant to material degradation. The shell makes the metal nanoparticle inert to chemical reagents. Nano-sized objects which have only the shell and are devoid of the core are called nanoshells. They are also equivalently known as nanocapsules and nanobubbles. Nanoshells are concentric nanoparticles consisting of two components: a dielectric core and a metallic shell. These particles can be designed to have unique optical properties by altering the geometry. Due to the shell structure of these nanoparticles, the optical properties of the particle are extremely sensitive to the core to outer shell ratio. Geometry and material properties of nanoshells can be designed such that they are useful for biological applications. Nanoshells can be tuned to possess certain optical properties. Each of these vials of nanoparticles has been tuned to have different absorption spectra over the visible range. There are some other types of nanoshells which are made of metals such as gold and silver and have a dielectric core. Nanoparticles of metallic, semiconducting, and magnetic materials have recently generated interest in terms of research, because of their potential uses in optoelectronics, reprography, catalysis, chemical sensing, and biological sensing. Among the metallic particles, the study of colloidal gold particles particularly stands out, and it is one of the most widely studied systems. This nanostructure has a unique optical property in that by changing the relative sizes of the core and the shell, its surface plasmon resonance (SPR) can be tuned in a broad spectrum of wavelength. Researchers have developed a procedure to make gold nanoshells on silica treated with aminopropyltriethoxysilane (APES). They have also tuned the properties of this system in such a way that it has been used for biological imaging and the therapy of cancer cells. There are reports of gold shells with polystyrene cores. These gold nanoparticle decorated silica cores can be modified for further stability by using self-assembled monolayers of alkane thiols. Other kinds of nanospheres are formed by the deposition of silica on biological systems such as liposomes (Pradeep, 2009). Oxide nanoshells have diameters in the range of 10–20 nm. This can be changed depending upon the required size. In the case of metal nanoshells, like gold with silica shell, the thickness of the shell can be up to around 20 nm. It has been found that these nanoshells are highly porous in nature. One of the main uses of these hollow silica shells is in the form of containers of drugs. The outer surface of these shells can be used for attaching antibodies so that the silica shell antibody complex can be used to bind to a specific antigen in fluid systems.

8.2 TYPES OF NANOSHELLS

Nanoshells can be obtained by different methods, based on the size and the intended use of the nanoshell. When synthesizing nanoshells, it is important to achieve controlled and uniform coating of core particles with the shell material. Several methods of synthesizing nanoshells have been established, but controlling the thickness of shell material coating on the core particles is difficult. These methods for synthesizing nanoshells suffer from the disadvantage of nonuniform coating. A favorable method to obtain desired coating involves surface precipitation of inorganic molecular precursors on particles, and removal of the core by thermal procedure, depending on the type of the shell (Pradeep, 2009). There are two types of nanoshells: oxide nanoshells and metal nanoshells. Fig. 8.1 illustrates different types of core–shell particles.

8.2.1 OXIDE NANOSHELLS

Oxide nanoshells are formed from the oxide core–shell particles with a hollow core. This group includes silica, titania, and zirconia nanoshells. One of the unique features of oxide nanoshells is that they have a hollow core and a covering made of oxide. The most important application of this type of nanoshell is in the field of encapsulation of molecules and spectroscopy.

8.2.1.1 Hollow silica nanoshells

Silica provides a few advantages for use as a protecting agent. Silica is chemically unreactive, so it does not interact with the reactions that take place within the core of nanoshells. It is optically transparent, which allows us to study the spectroscopy of the nanoshell systems easily. These silica nanoshell systems are useful in the study of the photochemistry of molecules and fluorophores (Pradeep, 2009).

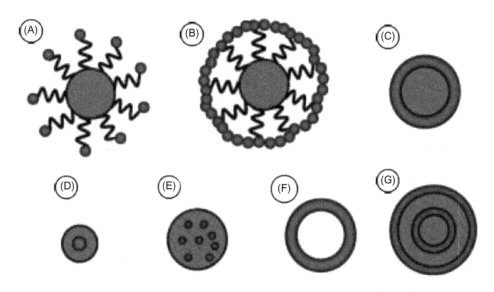

FIGURE 8.1

Variety of core–shell particles. (A) Surface-modified core particles anchored with shell particles. (B) More shell particles reduced on to core to form a complete shell. (C) Smooth coating of dielectric core with shell. (D) Encapsulation of very small particles with dielectric material. (E) Embedding number of small particles inside a single dielectric particle. (F) Quantum bubble. (G) Multishell particle.

From Kalele, S., Gosavi, S., Urban, J., & Kulkarni, K. (2006). Nanoshell particles: synthesis, properties and applications. Current Science, 15.

There are several steps required for the synthesis of silica-covered gold core–shell particles. The steps are the following:

1. Gold colloids are prepared.
2. 3-Aminopropyltrimethoxysilane (APS) is used as the primer and stabilizer to make gold surface vitreophilic, allowing for a thin layer of silica to form on gold colloid core.
3. Another precursor of silica, sodium silicate solution, is added at the appropriate pH to get a thicker layer of shell using the Stöber process.

The Stöber process is used to make large particles of silica with desired properties. This process allows formation of controlled silica particles within the range of 500 nm to 2 μm. The chemicals used for this reaction are tetraethoxysilane (TEOS) as silica precursor, water, ethanol, and ammonia. The silica particles obtained from the reaction mixture have a small size distribution and can be controlled by adjusting the pH of the solution, the composition of reactants, and the temperature. This process is currently used in various manufacturing processes. The hydrolysis of TEOS with water is very slow and ammonia is used as a catalyst. Hydrolysis promotes the formation of gel structures and ammonia is a morphological catalyst, which produces spherical particles. The reactions involved in the Stöber process are (Kalele, Gosavi, Urban, & Kulkarni, 2006):

$$Si(OC_2H_5)_4 + 4H_2O \rightarrow Si(OH)_4 + 4C_2H_5OH \qquad (8.1)$$

$$Si(OH)_4 \rightarrow (\text{in presence of } NH_3) SiO_2 + 2H_2O \qquad (8.2)$$

The first reaction shows the hydrolysis of TEOS and the second reaction shows the condensation of silica. Both reactions are base catalyzed. These reactions give the particles a negative charge, which stabilizes the surface. This method can be used to synthesize silica-covered gold and silver core–shell particles.

Procedures to make gold particles with definite size are available, and it is thus possible to obtain core–shell particles with a fixed core size. Using the core–shell particles, nanoshells can be attained by removing the core material. This requires using a suitable procedure to remove the core that does not affect the shell structure. One of the most used procedures for this application is using a cyanide ion for gold particles and ammonia for silver particles. To get nanoshells from silica-covered gold core shells, sodium cyanide is added to the solution, which dissolves the core of the particle (Pradeep, 2009). The dissolution of the core is observed using absorption spectroscopy from the disappearance of the surface plasmon peak of the gold nanoparticle. The reaction is shown below:

$$4Au + 8NaCN + 2H_2O + O_2 \rightarrow 4NaOH + 4NaAu(CN)_2 \qquad (8.3)$$

8.2.2 METAL NANOSHELLS

Metal nanoshells have a dielectric core composed of silica, which is different from oxide nanoshells in terms of structure. The structure of metal nanoshells provides a significant of optical properties that be adjusted based on the thickness of the shell. Gold nanoshells are part of metal nanoshells that provides opportunities for applications in the field of cancer treatment (Pradeep, 2009).

8.2.2.1 Gold nanoshells

One type of metal nanoshells is gold nanoshells, which are very useful in the field of cancer detection and treatment. The properties of gold nanoshells can be adjusted to scatter or adsorb light in a broad spectral range, nearly including infrared (IR). Near IR (NIR) is a wavelength region that provides maximum penetration of light through the tissue. This allows designing nanoshells that can be used for therapeutic and diagnostic applications. Gold nanoshells provide several advantages over silica nanoshells with respect to optical properties and absorption. The optical properties and absorption of gold nanoshells can be altered based on the thickness of the gold layer on silica core particles. The gold surface of the nanoshells provides the advantage of attaching different biomolecules. The gold surface is biocompatible, so it does not present any challenges to the functionality of the body. By using polyethylene glycol (PEG), the gold surface can be tailored to attach the desired type of molecules (Pradeep, 2009) (Fig. 8.2).

The following procedure is used to synthesize nanoshells:

1. Using the Stöber method, 100 nm diameter silica nanoparticles are prepared.
2. The surface silica particles are functionalized using APTES.
3. Small gold colloids are grown using the Duff and Baiker method and are adsorbed on silica nanoparticles.
4. More gold is grown into the nucleation sites using potassium carbonate and formaldehyde (Fig. 8.3).

8.2.2.2 Silver nanoshells

Silver nanoshells on silica nanoparticles can be prepared by a seed growth approach. This method allows us to get the desired optical properties of the silver nanoshells by adjusting the thickness of the silver shell. The Mie resonance of silver nanoshells takes place at energies different from any bulk interband transition. This allows silver colloid to have a stronger and sharper plasmon resonance than

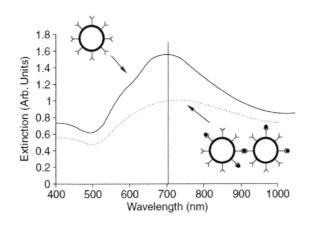

FIGURE 8.2

Gold nanoshell spectral range. UV–visible spectrum of dispersed nanoshells fabricated with 96 nm diameter core and 22 nm thick gold shells.

From Pradeep, T. (2009). Nano the essentials: Understanding nanoscience and nanotechnology. New Delhi: Tata McGraw-Hill Publishing Limited Company.

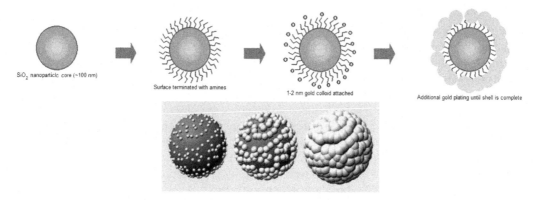

FIGURE 8.3

Synthesis of gold nanoshell having silica as core.

With permission from Naomi Halas - Optics and Photonics News (2002). Vol. 13, Issue 8, pp. 26–30.

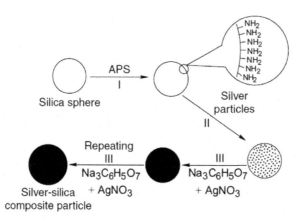

FIGURE 8.4

Fabrication procedure of silver nanoshells. As described in the text, using a silica sphere and a salinizing agent, silver particles are grown and adjusted as desired.

From Pradeep, T. (2009). Nano the essentials: Understanding nanoscience and nanotechnology. New Delhi: Tata McGraw-Hill Publishing Limited Company.

gold. A significant advantage of silver colloid is that the plasmon resonance of a solid silver nanoparticle appears at a shorter wavelength than of gold nanoshells (Pradeep, 2009).

The following procedure is used to synthesize nanoshells:

1. Silica sphere core is treated with amine-terminated surface silanizing agent.
2. The terminated amine groups are used as attachment points from small colloidal silver particles. They are used for growth of a silver nanoshell overlayer.
3. Silver particles are grown using the standard sodium citrate route.

Step 3 is used to adjust the thickness of the silver shell as required (Fig. 8.4).

8.3 **PROPERTIES OF NANOSHELL PARTICLES**

The coating of colloidal particles with shells offers a simple and flexible way of modifying their surface chemical, reactive, optical, magnetic, and catalytic properties. CdSe nanoparticles coated with CdS or ZnTe and CdTe nanoparticles coated with CdSe have been studied for enhancement in their luminescent properties. Semiconductor particles, such as ZnS doped with Mn, can be embedded inside a single silica particle. This doped semiconductor nanoparticles are known to be highly efficient fluorescent materials. The coating of dielectric materials such as silica can enhance the luminescence properties. Functional materials with novel properties can be synthesized using various combinations of core–shell materials and shell thickness. Below is a brief review of the properties of nanoshells (Kalele et al., 2006).

8.3.1 **OPTICAL PROPERTIES**

The metal nanoparticles show optical absorption in the visible range of the electromagnetic spectrum. The position of the absorption band shows small variations with particle size. The coating of metallic shells on silica allows one to tune the absorption band from the visible to the IR region. The relative thickness of core to shell layer is sensitive toward the position of the SPR band. Therefore, by changing the shell thickness, one can tune the SPR band position in the desired wavelength range as shown in Fig. 8.5 (Kalele et al., 2006).

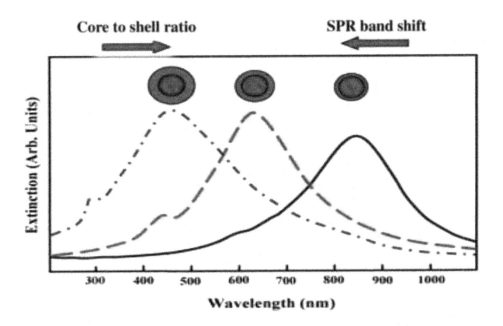

FIGURE 8.5

Variation in SPR band with shell thickness. By changing the shell thickness, it is possible to choose a position for the SPR band.

From Kalele, S., Gosavi, S., Urban, J., & Kulkarni, K. (2006). Nanoshell particles: synthesis, properties and applications. Current Science, 15.

The metal nanoshells that have plasmon resonance in the IR region are suitable for biological applications, as this electromagnetic range is transparent for biological tissues. Interaction between nanoparticles depends upon the separation between neighboring particles. Thick coating leads to larger separation of the metal particles, whereas thin coating leads to less separation. The dipole to dipole coupling between the particles is responsible for red shifting the plasmon band. If the particles are well separated (thick coating), the dipole to dipole coupling is fully hidden and the plasmon band is located almost at the same position as the individual metal particle. By varying the thickness of the shell by a small value, the color of the core–shell particles can be tuned from one color to another. At the same time, changes can be monitored spectroscopically by monitoring the SPR bands (Kalele et al., 2006).

8.3.2 LUMINESCENCE PROPERTIES

Semiconductor nanoparticles are fluorescent materials. The coating of silica is applied to them to decrease photo bleaching. The semiconductor nanoparticles coated with another layer of semiconductor have proven to be of great importance in enhancing the luminescence of these core–shell assemblies. The choice of shell material is important for localization of the electron hole pair. As in Fig. 8.6, in type I nanostructures such as CdSe/CdS or CdSe/ZnS, the conduction band of the shell material, which is a higher bandgap material, is at higher energy than the core, whereas the valence band of the shell is at lower energy than the core. In these materials, electrons and holes are confined in the core.

In type II nanostructures such as CdSe/ZnTe or CdTe/CdSe, both valence and conduction bands of the core material are at higher or lower energy than in the shell. In this case one carrier is confined in the core and the other in the shell. Type I and type II nanostructures have different properties because of the spatial separation of carriers. The lifetime decay of exciton and quantum yield of core–shell nanoparticles is much higher than individual semiconductor nanoparticles. The organic dyes are well-known

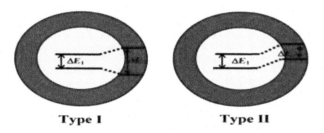

Type I   **Type II**

FIGURE 8.6

Type I and type II semiconductor core–shell structures. In type I nanostructures, the conduction band is at a higher energy than the core, while the valence band is at a lower energy. In type II nanostructures, both valence and conduction bands of the core material are at higher or lower energy than in the shell.

From Kalele, S., Gosavi, S., Urban, J., & Kulkarni, K. (2006). Nanoshell particles: synthesis, properties and applications. Current Science, 15.

phosphor materials which are similar to semiconductor nanoparticles. They are used as fluorescent biological labels. These dyes are not photostable, thereby quickly bleaching out. Furthermore, some of the dyes cannot be dispersed homogeneously in water. When these dye molecules are entrapped in silica shell, the enhancement of luminescence can be observed. The silica coating on these dye molecules makes them disperse uniformly in water (Kalele et al., 2006).

8.3.3 THERMAL PROPERTIES

The melting point of nanoparticles is lower than that of similar bulk material. This has been attributed to large surface tension in nanoparticles. In order to release this tension, it melts faster than bulk. Encapsulation of silica on these nanoparticles greatly improves the thermal stability of these particles. By changing the thickness of the shell, the variation in melting point can be observed. In some nanoshell assemblies such as metallic shells on dielectric cores, the thermal instabilities can be observed. Complete distortion of the shell can be observed when silica gold nanoshell particles are heated. The melting of nanoshells can be observed at significantly lower temperature. Due to higher surface area of core–shell particles, a higher number of particles is exposed to the surface and is affected by faster melting. When the particles are encapsulated with silica, the enhancement in thermal stability can be observed. It has been proven that 60–70 nm thick coating of silica greatly improves the thermal stability of gold nanoshells. The coating of silica on such shells is a way of preserving the identity of individual core particles because of the high temperature stability of silica (Kalele et al., 2006).

8.3.4 SURFACE CHEMICAL AND CATALYTIC PROPERTIES

The core–shell particles offer high surface area and can be used as efficient catalysts. Titania is a significant photocatalytic material. It has been established that nanoshells and nanoparticles show different catalytic behavior from bulk titania. Titania is thermally unstable and loses its surface area readily. Coating a thin layer of some other stable oxide such as silica on titania can greatly improve its catalytic activity (Kalele et al., 2006).

8.3.5 MAGNETIC PROPERTIES

Stability of magnetic materials is important when studying their magnetic properties. In order to improve the surface characteristics and protect them from reacting with various species to form oxides, they are coated with inert materials. Silica is a good choice because it forms stable dispersions. It is also nonmagnetic and therefore does not interfere with the magnetic properties of the core particles. Magnetic materials are often susceptible to agglomeration and show anisotropic interactions. Their stable dispersion can be prepared by inducing surface charges on them or adsorbing some organic molecules on their surfaces. Since organic molecules do not form any strong chemical bond such as covalent bond with magnetic particles they can be desorbed. When magnetic particles coated with silica are suspended in the medium, isotropic interactions are observed. Two magnetic materials can be used as core and shell. The magnetic properties can be tailored by varying core to shell dimensions (Kalele et al., 2006).

8.4 APPLICATIONS OF NANOSHELLS

This section will list and briefly explain significant applications of nanoshells. Nanoshells are often used for biomedical imaging, therapeutic applications, fluorescence enhancement of weak molecular emitters, surface-enhanced Raman spectroscopy, and surface-enhanced IR absorption spectroscopy.

8.4.1 PROTECTIVE WATERPROOF COATINGS FOR WOOD, METAL, AND STONE

Nanoshells are used to synthesize a range of coatings that can be applied to protect a wide variety of surfaces, such as stone, timber, metal, glass, and polished painted surfaces. The application of a coating can provide a long-lasting protection to materials, without affecting their texture or natural color.

8.4.2 ION-SELECTIVE FILMS

The layer-by-layer assembly of TiO_2 nanoshells or poly acrylic acid acts as an excellent detection tool for dopamine. Dopamine can be detected electrochemically by carbon fiber electrodes, but the interference from ascorbic acid hinders this process as it also falls in this electrochemical window. Nanoshells can be used without interference from ascorbic acid.

At a pH of 7, TiO_2 nanoshells are negatively charged and therefore, the diffuse part of the electrical double layer is composed primarily of cations, an ideal condition necessary for the selective detection of positively charged dopamine over negatively charged ascorbic acid. The ratio between the dopamine and ascorbic acid signals changes from 1:3 for a native glassy carbon electrode surface to 9:1 for a nanoshell-modified surface, which gives an overall 27-fold enhancement of the selectivity between these substances. The signal from the mixture of ascorbic acid and dopamine is virtually equal to that from 1 mM dopamine. The ascorbic acid peak current is negligible under these conditions (Pradeep, 2009).

8.4.3 GOLD NANOSHELLS FOR BLOOD IMMUNOASSAY AND CANCER DETECTION AND THERAPY

In the immunoassay procedure proposed by Halas and West, nanoshells are conjugated with antibodies that act as recognition sites for a specific analyte. The analyte causes the formation of dimmers, which modify the plasmon-related absorption feature in a known way. A fast absorption measurement can determine the presence of the molecule, avoiding the purification step.

Since these nanoshells have a large optical scattering cross section, they can be used as potential contrasting agents for photonics-based imaging modalities. Among the methods used, reflectance confocal microscopy and optical coherence tomography, which facilitate early cancer detection, are important. Optical properties of the nanoshells can be tuned in such a way that that they can be used for both imaging and therapy. The colloidal regime allows controlling both scattering and absorption properties simultaneously by changing size. Selective accumulation of the nanoshell can be used to image the tumor by using the high permeability and retention properties of the cancer cells (Pradeep, 2009).

HER2 is a protein whose acronym stands for human epidermal growth factor receptor 2. HER2 receptors are over-expressed in the case of a cancer cell. The use of nanoshells for therapy can target and kill the cancer cells without harming the surrounding healthy cells (Pradeep, 2009). Fig. 8.7 shows the application of nanoshells for cancer therapy.

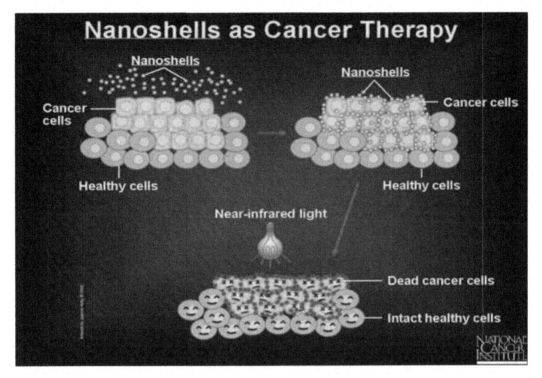

FIGURE 8.7

Nanoshells for cancer therapy. Nanoshells can be used to kill cancer cells without harming healthy ones.

From National Cancer Society (2008).

8.5 CURRENT RESEARCH AND FUTURE PERSPECTIVES

As evident from this chapter, nanoshells are important in this emerging field of nanotechnology. For this reason, much research is continuously being done on their fabrication and application. One of these heavily researched fields is the preparation of *metal* nanoshells. In this chapter, the common methods for synthesizing such nanoshells have been described. They generally involve binding metal colloidal nanoparticles to silane agent-coated silica particles. There is much research dedicated to finding alternative methods for making nanoshells. For example, Zhang et al. (2004) have prepared silver nanoshells without using a silane agent or colloidal particles. After preparing silica particles by a modified Stöber process, they made the nanoshells through a two-step process, where they first concentrated the reaction precursor, $AgNO_3$, on the surface of the particles and then added formaldehyde for reduction. Fig. 8.8 shows the results from their work.

Another interesting method for synthesizing metal nanoshells is the use of biological machinery. Viruses are particularly interesting, because they come in a variety of morphologies, such as spheres and rods, and have surfaces that can be used to work with a variety of chemistries. Furthermore, the capsid of a virus, the protein shell surrounding the genetic material, can be modified in a variety of

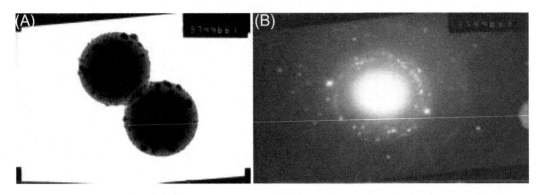

FIGURE 8.8

(A) TEM image of the silica particles with silver nanoshells. (B) Electron diffraction pattern of the silver nanoshells on silica particles.

With permission from Naomi Halas - Optics and Photonics News (2002). Vol. 13, Issue 8, pp. 26–30.

ways to one's advantage. Radloff et al. were able to assemble gold nanoshells on virus "bioscaffolds." They attached gold nanoparticles to the surface of the Chilo iridescent virus (CIV) and through little chemical modification were able to create biotemplates with the virus as the dielectric core. Fig. 8.9 shows the gold nanoshell growth on the CIV. With a wide variety of capsid architecture and their convenient chemical functionality, it is possible to create nanoshells with properties that are not feasible with silica particles. One such example is the size of the diameter, where nanoshells with smaller diameters may be easier to create using viruses.

Understanding the properties of nanoshells and how to manipulate is very important, and much research is dedicated to doing so. Nanoshell optical studies are very common because of the wide variety of potential technological applications. In order to study such properties, Averitt, Westcott, and Halas (1998) performed ultrafast optical studies on gold nanoshells. They manipulated the surface resonance by changing the ratio of the core diameter to shell thickness and observed transient bleaching and absorption for the gold nanoshells embedded in polyvinyl alcohol. They found that the electron dynamics in the thin gold shells are related to the transmission change of the nanoshell films. Furthermore, in determining the optical response, they found that even when the plasmon frequency is shifted below the onset of interband transitions, interband effects are important for determining the nonlinear optical response.

Particularly interesting for its potential in many applications is the ability to shift peak plasmon resonance wavelength into the NIR. It can be used in sensing and monitoring, astronomy, and many biomedical and optical applications. Oldenburg et al. (1999) studied the IR extinction properties of nanoshells. They too controlled the surface resonance by changing the ratio of the core diameter to shell thickness, resulting in nanostructures with an optical extinction that can be varied from 800 to 2200 nm. They found that the largest cores with the thinnest shells result in the largest frequency shifts into the IR. Furthermore, the low order aggregation of the nanostructures results in resonances placed further into the IR. The ability to manipulate nanoshells to vary the extinction properties may prove very useful in making certain devices and components depending on such wavelengths.

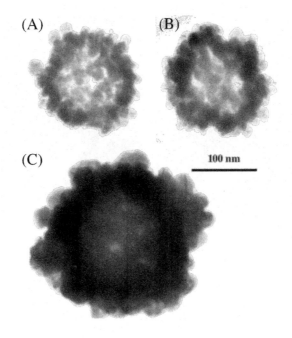

FIGURE 8.9

TEM images of gold nanoshell growth on the virus. The images show a clear progression of nanoshell growth as the nanoshells further coalesce from (A) to (C).

From Radloff, C., & Vaia, R. (2005). Metal nanoshell assembly on a virus bioscaffold. Nano Letters, 5.

One of the biggest and most promising applications is in the fight against cancer. An example is the use of nanoshells and specific wavelengths for tumor therapy. Hirsch et al. (2003) manipulated gold nanoshells such that they have strong absorption in the near IR, where maximal penetration of light through tissue is observed. The nanoshells are adsorbed or inserted into tumor cells and are then treated with NIR. This light heats the localized tumor containing the shells and kills them. This was done both in vitro and in vivo, and photothermally induced death of the cells was successful in both. As shown in Fig. 8.10, using both nanoshells and NIR results in cell death, while using either one alone does not. This method may prove very useful because it only results in the death of the specific cells with which the nanoshells are incorporated, thus leaving the healthy cells alive.

It is possible to incorporate these nanoshells into liposomes, which are artificially prepared vesicles, prior to photothermal therapy. They may also be used for drug delivery systems. Desai et al. set out to fabricate and characterize two popular gold nanoshells, gold–gold sulfide nanoshells and silica–gold nanoshells, which may be of further use to such applications. They synthesized gold–gold sulfide nanoshells by reducing aurochloric acid to gold sulfide using sodium sulfide and then adding sodium sulfide again to produce the nanoshells. The synthesis of silica–gold nanoshells was carried out similar to methods discussed throughout this chapter, where the fabrication begins with the Stöber process, and then the silica nanoparticles are surface functionalized, attached with gold nanoparticles, and reduced

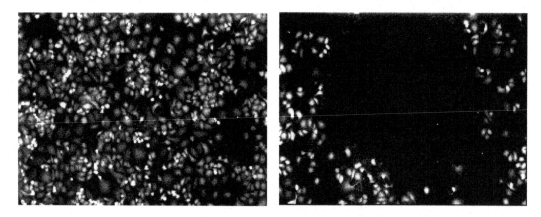

FIGURE 8.10

Images of tumor cells. The image on the left was taken only after the cells were exposed to NIR (without the nanoshells). The cells in the image on the right were given the nanoshells and were treated with NIR. The gap or dark spot in the right image is where the tumor used to be, before dying due to this therapy.

From Hirsch, L. R., Sershen, S. R, Halas, N. J., Stafford, R. J., Hazle, J., West, J. L. (2003). Nanoshell mediated near infrared photothermal tumor therapy. Department of Bioengineering, 4.

with aurochloric acid to form a layer of gold. Desai et al. synthesized particles "well below 100 nm which is ideal for drug delivery applications."

In cancer treatment, it is also important to be able to diagnose and detect the cancer cells. Loo, Lowery, Halas, West, and Drezek (2005) researched combined imaging and therapy of breast cancer cells using HER2-targeted nanoshells. The *HER2* gene has a significant role in cell growth and division. The increase in HER2 receptor is often responsible for breast cancer. As of right now, current molecular imaging techniques used to detect cancer does not provide a way to treat the cancer cells. Loo et al., however, designed the nanoshells to look for breast cancer biomarkers on the surface of cancer cells. In order to accumulate nanoshells in cancer tumor, they used immunotargeted nanoshells. In this study, the nanoshells were engineered, characterized, and attached by PEG linker to a nonspecific or anti-HER2 antibody. When the nanoshell bioconjugate is injected in the body, it accumulates in cells that overproduce proteins targeted by the attached antibody. The ability of the nanoparticles to absorb NIR light is used to photothermally destroy cancer cells. Fig. 8.11 shows that cells incubated with anti-HER2 nanoshells displayed a significant increase in scatter-based optical contrast, and cell death was only observed in cells exposed to anti-HER2 nanoshells and NIR laser.

Biological reactions move very quickly and are thus often very hard to detect. Improving sensors such that they can provide effective detection for a variety of applications (certainly not limited to this example) is an important field of research. For more information on the applications of sensors, read Chapter 4, Nanosensors. For example, in one study, Raschke et al. (2004) explained that molecular sensors based on single gold nanoparticle light scattering can be improved by the use of gold nanoshells, instead of solid gold nanospheres. The improvement is threefold. First, the particle plasmon resonance of nanoshells appears at lower energies compared to the plasmon resonance of nanospheres of the same diameter. The second advantage is that the particle plasmon of nanoshells exhibits a larger spectral shift than the particle plasmon of nanospheres, provided that they experience the same change

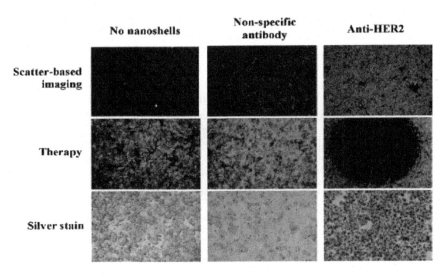

FIGURE 8.11

Combined imaging and therapy of breast cancer cells using HER2-targeted nanoshells. The top row shows scatter-based dark-field imaging of HER2 expression. The middle row shows calcein staining imaging. The bottom row shows silver stain assessment of nanoshell binding. The method evidently works because of the increased contrast seen in the top rightmost image and the cytotoxicity (dark spot) seen in the middle row rightmost image.

From Loo, C., Lowery, A., Halas, N., West, J., Drezek, R. (2005). Immunotargeted nanoshells for integrated cancer imaging and therapy. Departments of Bioengineering and Electrical and Computer Engineering, Rice University, Texas.

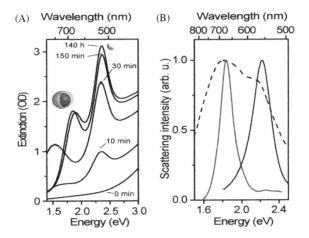

FIGURE 8.12

(A) Extinction spectra measured during the synthesis of Au_2S/Au nanoshells. (B) Scattering spectra of three individual nanoparticles: A Au_2S/Au nanoshell (red line (gray line in print version)) and solid nanospheres with diameter of 40 nm (solid black line) and 150 nm (dashed black line).

From Raschke, G., et al. (2004). Gold nanoshells improve single nanoparticle molecular sensors. Nano Letters, 5.

of refractive index in their direct vicinity. Finally, the third advantage is the much smaller full width at half-maximum of the scattering spectrum. Fig. 8.12A is the energy versus extinction graph for the wavelength region 400–800 nm, taken directly from the study. It shows that the gold nanoshells appear at lower energies than nanospheres of the same diameter. Fig. 8.12B is the energy versus scattering intensity graph for the wavelength region 500–800 nm. It shows that gold nanoshells exhibit a larger spectral shift than nanospheres. The study concluded that the gold nanoshells have the advantage of a narrow nanoshells particle plasmon resonance in the biological window of the optical spectrum and they show a larger plasmon shift for the same amount of change in refractive index of the surrounding nano environment compared to solid gold nanospheres.

ELECTRONIC AND ELECTRO-OPTIC NANOTECHNOLOGY*

9

9.1 HISTORY

The widespread interest in nanotechnology lays partly in the recognition that nanoscale materials (generally accepted to be on the order of 100 nm) exhibit novel properties and behavior (Poole & Owens, 2003). Properties such as conductivity and electronic structure can be altered, which is of particular interest to those who wish to explore the electronic and electro-optic potential of nanomaterials.

The systematic study of nanomaterials can be traced back as far as 1857, to the experimentation of Michael Faraday (Shah & Ahmad, 2010). It was Feynman in 1960, however, who delivered a speech to the American Physical Society which is seen as the seed of modern nanotechnology, with many of his proposed technologies becoming reality (Poole & Owens, 2003). By the early 1970s, researchers at Bell and IBM were developing techniques to build semiconductor layers atom by atom (Poole & Owens, 2003).

The discovery and manipulation of fullerenes, an entirely carbon molecule, allowed science to further explore the molecule's potentials in nanotechnology. This then led to the discovery of carbon nanotubes which became very useful in electronic and optic applications. Semiconductors are one example of carbon nanotube's application into nanoelectronics (Postma, Teepen, Yao, & Grifoni, 2001).

The synthesis and properties of semiconductor nanocrystals were developed soon after and electronic and optic devices became much faster. Other developments such as the scanning tunneling microscope also helped further nanoelectronic technology. This equipment gave researchers the ability to see and control atoms and be able to work on the nanoscale. Researcher Don Eigler in 1989 took advantage of this technology and was able to using the microscope to arrange 35 xenon atoms to spell out the IBM logo (Karkare, 2008).

Another major historical event in nanoelectronics is the discovery of self-assembly nanomaterials such as electrostatic self-assembly (Karkare, 2008). This technology was found to produce thin film materials with nanoscale-level molecular uniformity.

With the increasing demand for smaller, faster, and more highly integrated optical and electronic devices, as well as extremely sensitive detectors for biomedical and environmental applications, nanooptics or nanophotonics has become an emerging field, studying the many promising optical properties of nanostructures.

Nanophotonics is seen as a vital technology for extending Moore's law into the next few decades. In the past few years nanophotonics researchers worldwide have developed, On-chip Silicon Lasers,

---

*By Yaser Dahman, Dang Le, Natasha Niznik, and Niroshitha Sadyathasan.

Nanotechnology and Functional Materials for Engineers. DOI: http://dx.doi.org/10.1016/B978-0-323-51256-5.00009-5

Gigahertz Silicon Electro-Optic Switches, and Low Loss Highly Integratable Compact Nanowires (with waveguides of 100s of nanometers width) (Novotny & Hect, 2006).

Like nanotechnology itself, it is a rapidly evolving field. Because of the strong research activity in optical communication and related devices combined with the intensive work on nanotechnology, the field of nanooptics appears to be one with a promising future.

Nanophotonics is mainly expected to play a complementary role to micro/nanoelectronics on chip and extend the capacity of telecommunication networks into the terabit per second transfer rates (Novotny & Hect, 2006). One of the major emphases in the last few years has been developing on-chip interconnects to break the bottle neck for higher data rates within integrated chips.

Nanophotonic components such as microcavities with ultrahigh lifetime of trapped photons are expected to find applications in fundamental experimental physics such as gravitational wave detection (Fushman et al., 2008).

Intel, IBM, Lucent, and Luxera have highly functional and well-funded nanophotonic research groups. A number of universities in the United States, the United Kingdom, Japan, Italy, China, Belgium, etc. have been actively pursuing nanophotonics. Apart from a growing number of hits in publication databases like *Web of Science*, which shows it is already getting increased attention, it is also increasingly mentioned in the aims of the funding agencies, which will surely add to the activity in the field as increased economic support becomes available (Novotny & Hect, 2006).

9.2 THEORY OF ELECTRONIC NANOTECHNOLOGY

9.2.1 EFFECT OF SIZE ON METALS

When the sizes of metals reach the nanoscale, the distance between the particles reach such small sizes that they are on the order of the wavelength of an electron. This creates new electronic properties which are described by the Heisenberg uncertainty principle. In effect, the chemical properties of the metal become insignificant as compared to the effects caused by the size of the particle (Poole & Owens, 2003).

9.2.2 QUANTUM NANOSTRUCTURES

The classification of quantum nanostructures is dependent on how many dimensions the material is on the nanoscale. If the material is on the nanoscale in one dimension it is referred to as a quantum film or well. In two dimensions, it is a quantum wire. If it is on the nanoscale in all three dimensions, it is called a quantum dot (QD) (Shah & Ahmad, 2010). The directions in which charge carriers' movement is confined increase as the number of dimensions in the nanoscale increases, up to the QD in which the charge is confined to a small space, effectively zero dimensions of movement (Shah & Ahmad, 2010).

Quantum nanostructures can be prepared by both bottom–up or top–down approaches, such as lithograph (Poole & Owens, 2003).

QDs have several applications. One of the first applications found was their ability to emit very specific wavelengths of light. This is different from other light-emitting bulbs since QDs could be tuned across the visible and ultraviolet spectrums very precisely. Researchers have found that if they put about 2000 QDs together, they would have a finely tuned LED. Researchers have tried for an extremely long time to get these dots to emit light. In the 1990s, researchers were able to get a dark red light. Since then, other researchers have been able to tune the dots to a higher frequency, thus gaining blue and green light. The applications for this would be beneficial so that we could make full color screens and monitors (Wendy et al., 1999).

9.3 THEORY OF ELECTRO-OPTIC NANOTECHNOLOGY

On the topic of the optical properties of nanomaterials, Shah and Ahmad (2010) have the following to say:

> The linear and non-linear optical properties of such materials can be finely tailored by controlling the crystal dimensions and the chemistry of their surfaces. Fabrication technology becomes a key factor for the applications.

This effectively encompasses some of the challenges in electro-optic nanotechnology, including those which are presented in the literature review.

9.3.1 ABSORPTION AND EMISSION TRENDS

The absorption and emission of light in nanomaterials is an important consideration in the production of displays, particularly those of organic light-emitting diodes (OLEDs).

One interesting property of semiconducting materials on the nanoscale is that they have optical properties which differ from the bulk material. It has been noted that as the size of the particles is reduced, there is a shift in the absorption spectra to the blue (Poole & Owens, 2003).

Once again, we see that size and structure play a greater role than chemical identity in this property. Nanocrystals with larger bandgaps emit lower wavelength colors such as blue and green, whereas low bandgap materials emit higher wavelength colors like red (Shah & Ahmad, 2010).

9.3.2 DISPLAYS

In liquid-crystal displays, the resolution, brightness, and contrast are all dependent on the grain size of the particles being used. It is for this reason that nanomaterials are being investigated for use in such applications (Shah & Ahmad, 2010).

9.3.3 POLYMER-DISPERSED LIQUID CRYSTALS

The polymer-dispersed liquid crystals (PDLCs) are materials on a nano-level that can be thought of as Swiss cheese; there is a sold structure with scattered fragments in between. The cheese itself is a polymer substance which has tiny holes that are filled with liquid crystals. Liquid crystals have properties which are associated with both crystals and liquids (Chandrashekar, 1992).

The liquid-crystalline materials are the most crucial part of creating the PDLCs. There needs to be a perfect balance of rigid parts to help align the molecules in one direction and flexible parts to ensure there is fluidity in the liquid crystal (Bryant, 2001).

There are three major production methods of PDLCs. The first method is encapsulation, also known as emulsification. This method requires that the liquid crystal, polymer, and water be mixed in a solution. Once thoroughly mixed, the water is evaporated leaving the liquid crystals trapped in the polymer layer. The second method of producing PDLCs is phase separation. There are two different types of phase separation, polymerization-induced phase separation and thermally induced phase separation (Malik, Bubnov, & Raina, 2008). The polymerization-induced phase separation has three major steps. The first is to mix a liquid prepolymer solution and a liquid crystals solution. This solution is mixed into a homogenous state and then the activation is started to form polymers from the prepolymer solution capturing the liquid crystals in between the polymers as it forms. The thermally induced phase separation also has three major steps. The first is to obtain a polymer solution and heat it until the binds of the polymer loosen. Following this, liquid crystals need to be mixed with the heated polymer solution and then mixed to a homogenous state. The solution is then allowed to cool and the polymer will then obtain the liquid crystals that have seeped into the loosened binds.

This material has been paired with electricity to create smart windows (Bryant, 2001). The electrical conducting properties that have been incorporated into the liquid crystals on a nanoscale have allowed the crystals to be controlled by an electric volt. The amount of electricity applied to the liquid crystal will cause a rearrangement on the nanoscale of its structure causing it to scatter or become aligned within the polymer layer. The natural state of a smart window appears opaque due to the random orientations of the liquid crystals in the polymer. This state helps to block out sunlight saving energy on air conditioning, because the sunlight can be deflected causing less heat into rooms (Oltean, 2006). Once an electrical signal is sent by a controlling device, the scattered liquid crystals align allowing the sunlight to be absorbed and the glass appears transparent. This state allows for more natural light saving on cost of lighting and also allows for natural heating for a room. The overall savings can amount to a maximum of 20% savings on electric cost (Oltean, 2006). Another added benefit from smart windows besides saving on electricity cost is that it allows the user privacy and the control of natural sunlight with a flip of a switch.

9.3.4 DENDRIMERS

Dendrimers are presented in this section because of their suitability for use in electro-optic materials (Cameron et al., 2002). Dendrimers are molecules with a branched mode of growth (Poole & Owens, 2003). Dendrimers can be functionalized with various groups, but for the purpose of electro-optics, light sensitive chromophores are of the greatest interest (Poole & Owens, 2003). Dendrimers can be easily removed from the reaction mixture because of their large size and therefore are easy to work with (Poole & Owens, 2003).

9.3.5 MODULATORS

An application of nanophotonics is the electro-optic modulators which are devices used to modulate or modify a beam of light. Currently they are mainly used in the information technology and telecommunications industries (e.g., fiber-optic cables). Nanoscale optical communication devices will have increased speed and efficiency, once they can be engineered and used. Nanosize electro-optic modulators will be an integral part of a nanoscale communications network (Vlasov, Green, & Xia, 2008).

9.3.6 **PHOTODETECTORS**

Photodetectors are electro-optic devices that respond to radiant energy. They are basically sensors of light or other electromagnetic energy. A sensor is an electronic device that converts one type of energy to another for various reasons. Nanoscale size photodetectors will be an integral part of a theoretical nanoscale optical information network (Novotny & Hect, 2006). Nanotechnology creates many new, interesting fields and applications for photonic sensors. Existing uses, like digital cameras, can be enhanced because more "pixels" can be placed on a sensor than with existing technology. In addition, sensors can be fabricated on the nanoscale so that they will be of higher quality and possibly defect free. The end result would be that photos would be larger and more accurate. As part of a communication network, photonic sensors will be used to convert optical data (photons) into electricity (electrons). Nanoscale photonic sensors will be more efficient and will basically receive similar advantages to other materials constructed under the nanoscale (Novotny & Hect, 2006).

9.3.7 **ELECTROPHORETIC DISPLAY SYSTEMS**

The main goal of electrophoretic display systems is to produce a high-quality image for low power consumption (Ahn, Yu, Kim, Lee, & Kim, 2008). This display is generally made with titanium dioxide particles dispersed in hydrocarbon oil. The oil used in the display system has a dark-color dye added to it along with surfactants and charging agents. The surfactants will lower the surface tension and interfacial tension of the particles. This allows for easier movement of the particles through the oil allowing a faster transition time between displayed images (Das, Gates, Abdu, Rose, & Picconatto, 2007). The charging agents are added to allow particles to take on an electric charge and be more responsive even under low volts. Once this mixture has been formed and set between two layers of glass, an electric field can be used to rearrange the charged pigment particles to create visible images. This means that the dark colors will emerge to the viewers in that of the pattern fed into the electronic system, either revealing picture or letter which then can be changed via the electric field. This technology allows for thinner and lighter weight electronic reading devices with low power consumption.

9.3.8 **SWITCHES**

Electro-optic switches change signals in optical fibers to electrical signals. Typically semiconductor-based, their function depends on the change of refractive index with electric field. This feature makes them high-speed devices with low power consumption. Neither the electro-optic nor thermo-optic optical switches can match the insertion loss, back reflection, and long-term stability of opto-mechanical optical switches. The latest technology combines all-optical switches that can cross-connect fibers without translating the signal into the electrical domain. This greatly increases switching speed, allowing today's networks to increase data rates. However, this technology is only now in development and deployed systems cost much more than systems that use traditional opto-mechanical switches (Kimble, 2008).

9.3.9 **PHOTONIC CRYSTALS**

"Photonic crystals are composed of periodic dielectric or metallo-dielectric nanostructures that are designed to affect the propagation of electromagnetic waves (EM) in the same way as the periodic potential in a semiconductor crystal affects the electron motion by defining allowed and forbidden electronic energy bands. Simply put, photonic crystals contain regularly repeating internal regions of high and low dielectric constant" (Englund et al., 2007). Photonic crystals (PCs) are used to modify or control the flow of light. PCs may have a novel use in optical data transmission, but are not extremely prominent. They may be used to filter for interference in a fiber-optic cable or increase the quality of the transmission. In addition, they can be used to divide different wavelengths of light. PCs can already be manufactured at close to the nanoscale.

9.3.10 MULTIPLEXERS

A multiplexer is a device for converting many data streams into one single data stream, which is then divided into the separate data streams on the other side with a demultiplexer (Novotny & Hect, 2006). The main benefit is cost savings, since only one physical link will be needed, instead of many physical links. In nanooptics, multiplexers will have many applications. They can be used as part of a communication network, as well as utilized on a smaller scale for various modern scientific instruments.

9.4 CURRENT RESEARCH AND FUTURE PERSPECTIVES

There is no doubt that the field of electro-optic nanotechnology is one of the most valued areas in nanotechnology, because its applications are nearly limitless ranging from detectors to device displays.

Nonlinear optics is essential to the field of electro-optics. Swamy, Rajagopalan, Vippa, Thakur, and Sen (2007) present their results about a quadratic electro-optic effect in a nano-optical material. Quadratic electro-optic effect is a change in the refractive index of a material in response to an applied electric field. The nonconjugated polymer studied was poly(ethylenepyrrolediyl) (PEP) derivative with butyl substituent as shown in Fig. 9.1.

Conjugated polymers such as polydiacetylenes, often described as nano or quantum wires, are known to be exceptionally susceptible to large optical interference. These large susceptibilities are primarily due to the delocalization of π-electrons along the conjugated chain (quantum wire).

In contrast to conjugated polymers, nonconjugated polymers with isolated double bonds do not have delocalized electrons and are usually not expected to display significant nonlinear optical effects.

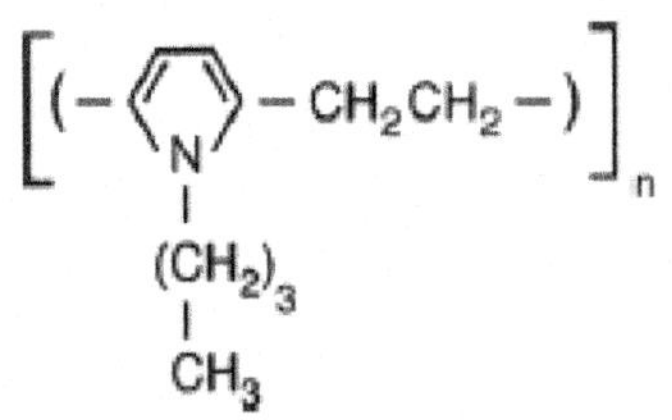

$$\left[(- \langle \rangle_{N} - CH_2CH_2 -) \right]_{n}$$
$$(CH_2)_3$$
$$CH_3$$

FIGURE 9.1

Molecular structure of PEP derivative with butyl substituent.

From Swamy, R., Rajagopalan, P., Vippa, P., Thakur, M., & Sen, A. (2007). Quadratic electro-optic effect in a nano-optical material based on the non-conjugated conductive polymer, poly(ethylenepyrrolediyl) derivative. Solid State Communications, 519–521.

Swamy et al. had layered a thin film of PEP on a glass slide. The film was about 0.3 μm in thickness. Fig. 9.2 shows the optical absorption spectrum of this polymer for different doping levels.

A He–Ne laser operating at a wavelength of 633 nm was used for the experiment. Metal (gold) electrodes with a gap were applied on the glass slide by evaporation prior to the deposition of the thin film of PEP. The modulation signal was also recorded using a lock-in amplifier (with 2f synchronization). The signal increased quadratically with the applied voltage. The signal as obtained for a field of 1 V/μm is shown in Fig. 9.3. The lower waveform represents the applied AC field at 4 kHz.

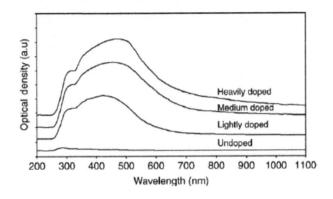

FIGURE 9.2

Optical absorption spectra of PEP derivative for different molar concentrations (y) of iodine. Lightly doped: y~0.3, medium doped: y~0.6, and heavily doped: y~1.0.

From Swamy, R., Rajagopalan, P., Vippa, P., Thakur, M., & Sen, A. (2007). Quadratic electro-optic effect in a nano-optical material based on the non-conjugated conductive polymer, poly(ethylenepyrrolediyl) derivative. Solid State Communications, 519–521.

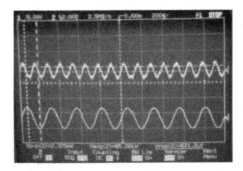

FIGURE 9.3

Oscilloscope trace of the modulation signal (top waveform) in quadratic electro-optic measurement of doped PEP.

From Swamy, R., Rajagopalan, P., Vippa, P., Thakur, M., & Sen, A. (2007). Quadratic electro-optic effect in a nano-optical material based on the non-conjugated conductive polymer, poly(ethylenepyrrolediyl) derivative. Solid State Communications, 519–521.

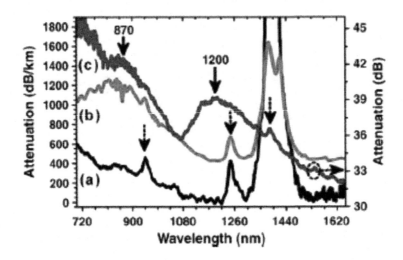

FIGURE 9.4

Optical attenuation spectra of (a) MOF 38 cm and (b) MOF 14 cm, both without PbS QDs, and (c) MOF containing a mixture of PbS877, PbS1160, and PbS1474 spread onto the core surfaces.

From Chillcce, E., & Ramos-Gonzales, R. (2010). Luminescence of PbS quantum dots spread on the core surface of a silica microstructured optical fiber. Journal of Non-Crystalline Solids, 2397–2401.

In summary, optical absorption, the quadratic electro-optic effect, and electroabsorption have been measured in the nonconjugated conductive polymer studied (iodine-doped PEP derivative), thus showing that doped PEP may have some future applications in nonlinear optics.

Broadband luminescence spectra, in the region from around 1000 to 1650 nm, of lead sulfide (PbS) QDs spread onto the surfaces of silica microstructured optical fibers (MOFs) (core diameters of 2.5 and 5.0 μm) have been reported. Chillcce and Ramos-Gonzales (2010) have injected and spread colloidal solutions of PbS QDs of different sizes with luminescence bands around 877 (PbS877), 1160 (PbS1160), and 1474 nm (PbS1474) onto the dual core surfaces of silica MOFs using a nitrogen gas pressure system. The PbS QDs were excited (via evanescent field effect) by the light of a continuum wave semiconductor laser or a Ti sapphire laser (at 785 nm) guided through the MOF cores. Fig. 9.4 is a graph of attenuation (lost in intensity of light) versus the wavelength of the laser. Curves a and b represent the MOFs of 38 and 14 cm, respectively, without PbS coating. Curve c is an MOF that is coated with a PbS mixture.

The PbS QDs luminescence spectra reveal blue-shift and band-broadening behaviors when increasing the pumping time duration as seen in Fig. 9.5A and B, which represent the MOF lengths of 38 and 14 cm, respectively. Broader and flatter luminescence spectra (from around 1000 to 1650 nm) were obtained using a mixture of QDs (PbS877, PbS1167, and PbS1474). By maximizing the signal amplitude, MOF with PbS QDs may be used as an active optical device with potential application in optical amplification and sensing.

Switching light in an effective way is important for optical networks (Li, Yu, & Yang, 2010). Li et al. (2010) demonstrated the possibility to realize an electro-optic switching operation in a weak-excitation

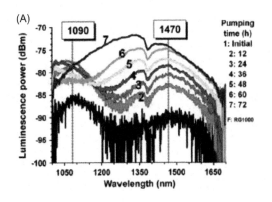 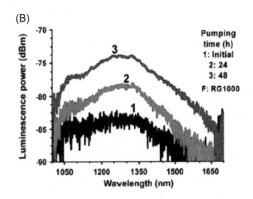

FIGURE 9.5

(A) A luminescence spectra of the MOF 38 cm length with mixed PbS QDs. (B) Luminescence spectra of the MOF 14 cm length with mixed PbS QDs.

From Chillcce, E., & Ramos-Gonzales, R. (2010). Luminescence of PbS quantum dots spread on the core surface of a silica microstructured optical fiber. Journal of Non-Crystalline Solids, 2397–2401.

limit through a model, where one can control the path of a photon propagating in a waveguide by varying the applied voltage to tune the electron tunnel coupling (quantum tunneling). Quantum tunneling refers to the quantum mechanical phenomenon where a particle tunnels through a barrier that it classically could not surmount because its total kinetic energy is lower than the potential energy of the barrier.

One of the methods for achieving optical switching is based on quantum coherence and interference in the light–matter interaction. In conventional media, the main drawback of this technique is the weakness of optical nonlinearities and the high power consumption of the switching operation.

The model that they described involves combining two-dimensional PC cavity and semiconductor QDs or quantum dot molecule (QDM). Fig. 9.6 describes a single QD switch, where a single-mode PC cavity containing an embedded three sub-band QDM is symmetrically coupled to two identical parallel waveguides a and b. As seen in the figure, there is a_{in} and b_{in} where signals may enter, but the signal may exit only through a_{out} or b_{out}. Making use of the two-dimensional properties of the PC, an assumption is made that an input signal is guided by the waveguide b in the crystal plane (Fig. 9.6). Other symbols in the figure are used for the model calculations.

As the author works the mathematical equations of the model, it is shown that when the angular velocity (ω) of the voltage applied is set to 0, the QDM becomes "transparent" allowing a signal to jump from waveguide b to waveguide a. Thus, the ω will change the QDM back to an "opaque" state blocking the signal from jumping waveguides.

OLEDs are a technology used in a variety of devices, such as certain monitors, TV screens, and cell phones. Coya et al. (2008) investigated the production of a thin film of solution-based dendrimers which have blue emission for this technology. Previously, linear chain polymers have been used, but the many limitations they suffer (mainly stemming from poor control over morphology) appear to be more easily controlled in dendrimic structures. In the past, thin films of dendrimers have been prepared by

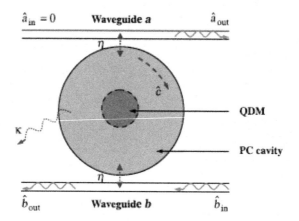

FIGURE 9.6

Schematic diagram of proposed electro-optic switching system, which is composed of one single-mode PC cavity, one three sub-band QDM and two identical parallel waveguides. The QDM is localized in the PC cavity.

From Li, J., Yu, R., Yang, X. (2010). Double-state controllable optical switching through three tunnel-coupled quantum dots inside waveguide coupled photonic crystal microcavity. Optics Communications.

evaporation. This does not provide the desired emission due to aggregation. Coya et al. prepare a film by spin coating, a method which has been previously used with polymeric materials.

The introduction of peripheral alkoxy chains has been shown to enhance the solubility in organic solvents and the electronic density of dendrimers. A comparison is made between dendrimers with and without this alkoxy ($-OC_6H_{13}$) addition.

A novel dendrimer is prepared from the reaction of triphosphate and *p*-hexyloxybenzaldehyde in dry tetrahydrofuran (THF) and tert-butoxide. This approach is chosen because of its good yield and all-*trans* configuration of the double bonds which remove the need for isomerization steps. The film is formed after the solution is kept in an ultrasonic bath and filtered through a 0.2 μm filter. It is then spin coated onto quartz substrates and cured.

It was found that substituted dendrimer suffers no aggregation based on its absorption and emission spectra. The suitability of the films formed using four different solvents is determined via both absorption and emission spectra, as well as visually with environmental scanning electron microscope (ESEM). Toluene had the effect of forming highly nonhomogenous films. $CHCl_3$ forms the most uniform film, followed by THF and C_6H_5Cl. The film produced by $CHCl_3$ had a thickness of 185 nm and a roughness of ±5 nm. $CHCl_3$ also has the least effect on shifting the maximum absorption and has the narrowest spectrum (Fig. 9.7).

It is concluded that the dendrimers containing the alkoxy chains result in a blue emission both in solution and in a thin film. Furthermore, the thin film is best produced using $CHCl_3$ as a solvent for spin coating. This method produces material suitable for the low-cost, large area, solution-processed OLED displays.

The electrophoretic display system has also been a subject of much interest. Ahn et al. (2008) explored the electro-optic response characteristics of a dual particle electrophoretic display system.

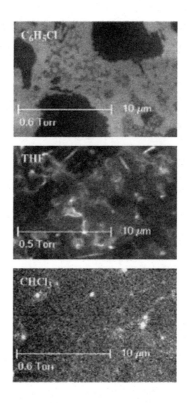

FIGURE 9.7

Environmental scanning electron microscope (ESEM) images of the obtained films.

From Coya, C., Álvarez, A. L., Ramos, M., de Andrés, A., Zaldo, C., Gómez, R., et al. (2008). A fluorescent stilbenoid dendrimer for solution-processed blue light emitting diodes. Organic Optoelectronics and Photonics III.

Electrophoretic display is often noted as electronic paper. These displays use electrophoresis to switch pixels on and off. The pixels are turned on and off by an electric field which floats the particles suspended in liquid to the visible surface. The two colored pigments used in their experiment are red and white. The dual-particle formulation allows for reduced thickness of the device and a faster response with lower voltage. They started the experiment by encapsulating the red pigment in polystyrene using an emulsion polymerization method. The red pigment was positively charged. The negatively charged white particle was made of rutile titanium dioxide. The interest in this study centered on finding the fastest response time of imaging with the lowest possible voltage amount. The red pigments had a mean diameter of 120 nm. The experiment expected that the particles would have high dispersion stability in the medium due to the strong repulsive Columbic interaction among the particles. When the bipolar electric field was applied, the white particles which were negatively charged moved to the positive electrode and the red particles, which were positively charged, move to the negative electrode. This was an expected result. However, the intensity was only 1.7 which was disappointing. Nevertheless, the experimenters did not have a problem identifying the image, so overall, the experiment was deemed successful. The potential problem that occurred in identifying intensity was that it was done using standards

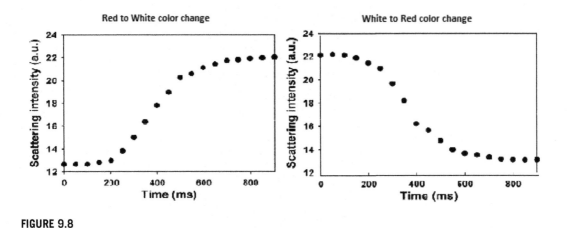

FIGURE 9.8

Color changes seen during experiment.

From Ahn, Y. H., Yu, G. C., Kim, Y. C., Lee, S.S., Kim, J. (2008). Electro-optic response characteristics of a dual particle electrophoretic display system. Molecular Crystals and Liquid Crystals, 492, 322–327.

used for liquid crystal displays. Perhaps using a new concept of color contrast ratio specifically for dual-particle color display would be more appropriate and accurate. The results however showed that the white to red color change was perfectly aligned as can be seen in Fig. 9.8.

The final findings of this experiment were that a response time for image change could be achieved in 650–750 ms and this could be done with a voltage of 1.4 V/μm. The future applications are to incorporate more than two colors and maintain the thin size of the device and the low power consumption.

PDLCs can too be used for displays in devices. They have a variety of applications within this area, and thus much research is dedicated to testing their effectiveness and attempting to improve them. No and Jeon (2009) examined if there is a benefit to aligning layers of PDLC displays to improve the electro-optic properties. A control cell where the PDLC is manufactured normally (with the scattered liquid crystals in the polymer material) is compared to a PDLC that has all of its liquid crystals in parallel aligned layers. The PDLCs were manufactured using the polymerization-induced phase separation. The mixture was placed in between two transparent indium tin oxide coated glass plates. The first observation made during this experiment is that aligned layered cells appeared to have liquid crystal droplets that were four to five times smaller than the control cell which can be observed in Fig. 9.9. This phenomenon is explained by the existence of the thin polymer layers having a nano-pattern which is formed during the rubbing process of creating thin layers to align. To ensure that the liquid crystals were aligned perfectly in parallel, a separate experiment was performed on the aligned layered PDLC. The experiment consisted of a laser light that was shot through the material and the refection of light was recorded. The light refracted showed a sinusoidal variation of transmittance of a polarized light with 180° period. This confirms that the liquid crystals were indeed aligned in parallel.

While testing the aligned layered cells to the control cells, there was no significant difference in performance quality. They both had similar contrast ratios and response times to varying voltages. The operating voltages in terms of temperature stability were also seen to be similar in both materials. The overall conclusion from this study is that even though alignment does make the liquid crystals more

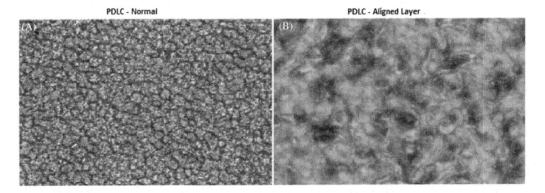

FIGURE 9.9

Comparison of liquid crystal droplets in treated and control cells.

From No, Y. S., Jeon, C.-W. (2009). Effect of alignment layer on electro-optic properties of polymer-dispersed liquid crystal displays.

Molecular Crystals and Liquid Crystals, 513, 98–105.

defined, the material does not perform any better. Thus, PDLC should be manufactured without alignment layers.

In another study focusing on PDLC, Malik et al. (2008) examined the electro-optic and thermo-optic properties of PDLC and guest–host polymer-dispersed liquid crystal (GHPDLC). The PDLC was manufactured using polymerization-induced phase separation. The investigation is centered on finding out if GHPDLC will result in the same and or even hopefully better results than that of PDLC. The GHPDLC is produced in the same way as the PDLC. However, a small amount of diachronic dye, which is noted as the guest, is placed into the liquid crystal host. This small change is seen to enhance the optical characteristic of the liquid crystalline material. The experiment requires that the following several factors (refractive index of the material, liquid crystal polymer concentration, and cooling rate) be kept constant for the purpose of the experiment. The focus will be on comparing the effect the dye has and how it affects the electric field and temperature. In this particular experiment, the PDLC is composed of a nematic liquid crystal and the GHPDLC is composed of nematic liquid crystal and a blue anthraquinone dichroic dye. During the experiment, the dichroic dye was ensured to have no ionic impurities and was seen to have a good solubility in the liquid crystal. Once both materials were ready, they were compared at varying voltages (0, 10, 30, and 100V) keeping frequency at 500Hz and at room temperature. The following images in Fig. 9.10 were taken at the varying voltages and show that the GHPDLC has an enhanced optical characteristic.

The results of the study proved that injecting a dye into the liquid crystal during the manufacturing maintained the characteristics of the PDLC and that in fact it improved the switch time.

Self-assembled monolayers (SAMs) are used in a wide variety of devices, attracting much research. Novak, Jager, Kropp, Clark, and Halik (2009) attempt to explain the unexpected current density behavior of long-chain *n*-alkyl phosphonic acids as compared to their shorter chain counterparts. Their behavior deviated from what was expected theoretically, as well as the trends set by shorter chain molecules of the same type. Specifically, they were investigating an SAM which is used in organic transistors that show promise in large area, thin electronic devices. They proposed that the behavior was caused by morphology and not by defects in the self-assembly. In order to investigate the behavior, they intended

PDLC GHPDLC

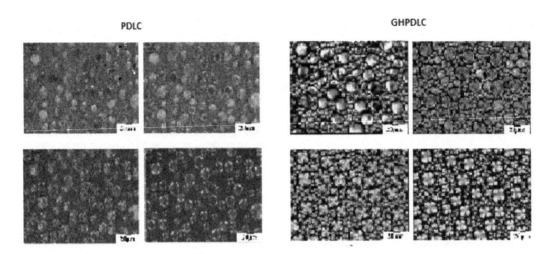

FIGURE 9.10

Comparison of PDLC and GHPDLC at varying voltages.

From Malik, P., Bubnov, A. M., & Raina, K. K. (2008). Electro-optic and thermo-optic properties of phase separated polymer dispersed liquid crystal films. Molecular Crystals and Liquid Crystals, 494, 242–251.

on both modelling the SAM on a computer and preforming experiments to determine the structure of the monolayer via conduction data.

The experimenters used p-doped silicon wafers as a substrate and grew an aluminum oxide film of about 2 nm by atomic layer deposition. They then generated the SAM from solution according to literature. Four different lengths of *n*-alkyl phosphonic acid were to be compared, C_6-, C_{10}-, C_{14}-, and C_{18}-PA. Capacitor stacks were then added (gold, in this case) via stencil mask. Electrical characterization of the transistors and capacitors was then carried out. Measurements were repeated to ensure accuracy. Simulation using the molecular dynamics program (NAMD) was performed in order to observe the interactions between the molecules based on a surface containing 80 phosphonic acids.

It was shown that the decrease in current densities as the chain length of the phosphonic acid increased was not exponential as expected. This indicated that the effective thickness of the layer was not as expected in the C_{14}-PA and C_{18}-PA chains. Computer simulations showed that the shorter length chains (C_6-PA and C_{10}-PA) formed a more amorphous solid, whereas the C_{14}-PA and C_{18}-PA chains had a more crystalline structure which featured gaps (Fig. 9.11).

It was proposed that it was the gaps which lead to a shorter effective thickness of the layer, because the thermally deposited gold can enter the gaps and allow for increasing the tunneling current in the monolayer. The van der Waals forces between the longer length chains appear to be the cause of the semicrystalline structure they form. It is noted by the authors that they do not know the effect the top layer has on the ordering of the molecules as they were not able to model that. It is suggested that these layers may even cause or increase gaps.

This discovery is of importance because it shows that even small changes in length can strongly effect the electrical characteristics of SAMs, which show great promise in molecular scale electronics.

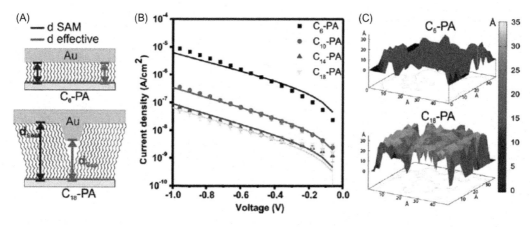

FIGURE 9.11

(A) Scheme of the proposed mechanism for C_6-PA and C_{18}-PA exemplarily. (B) Results of Simmons modeling and MD calculations. Fit of the current density data of capacitor devices (area: $150 \times 150\,\mu m^2$) for n-alkyl phosphonic acids. (C) Surface topography extracted from MD calculations for C_6-PA and C_{18}-PA.

From Novak, M., Jager, C., Kropp, H., Clark, T., Halik, M. (2010). The Morphology of Integrated Self-assembled Monolayers and Their Impact on Devices - A Computational and Experimental Approach. Organic Electronics Volume 11, Issue 8, August, Pages 1476–1482.

It is important to understand the effects morphology has as the device parameters are controlled via molecular design.

Bhattacharjee, Bhakat, Roy, and Kusz (2010) studied the electrical conduction properties of the molecular magnetic material $\{N(n\text{-}C_4H_9)_4[Fe(II)Fe(III)(C_2O_4)_3]\}\infty$ with three different cations. The goal was to determine the effect that the cation size had on the conduction of the material. In addition, the material was thermally degraded and the properties of the resultant material were determined. The three compounds to be tested are $\{N(n\text{-}C_4H_9)_4[Fe(II)Fe(III)(C_2O_4)_3]\}\infty$, $\{N(n\text{-}C_3H_7)_4[Fe(II)Fe(III)(C_2O_4)_3]\}\infty$, and $\{AS(C_6H_5)_4[Fe(II)Fe(III)(C_2O_4)_3]\}\infty$ and shall be referred to as BuFeFe, PrFeFe, and AsFeFe, respectively.

The compounds were prepared in a one-pot reaction and their conductivity was measured. It was determined that the compounds were found to be Ohmic under the conditions measured. It is suggested that the results fit in with the proposed hopping mechanism expressed by the Arrhenius relation, which measures the temperature dependence of the conductivity of materials. The materials are found to be semiconducting.

The effect of various unit cell lattice parameters on the conductivity is explored. It is shown that the values of these parameters follow PrFeFe< BuFeFe< AsFeFe and that the measured activation energy (part of the Arrhenius relation, and thus temperature dependence on conductivity) and the room temperature conductivity are directly related to these parameters, but not to a uniform extent. It follows that the larger the unit cell volume, the larger the activation energy and room temperature conductivity is in this set of materials. Magnetization of the materials does not follow this order, which suggests that a more subtle variation in the structure is responsible for the effect on this property. It is proposed that cations affect the metal oxide layer of the material and this affects the conduction. No clear relation

Table 9.1 Estimated Values of Room Temperature Conductivity σ_{RT}, Preexponential Factor σ_0, and Activation Energy for Cation Materials

Cation (Compound)	σ_{RT} ($10^{-15}\Omega^{-1}cm^{-1}$)	σ_0 ($\Omega^{-1}cm^{-1}$)	E (eV)	A_0 (Å)	C_0 (Å)
$N(n\text{-}C_3H_7)_4^+$(PrFeFe)	2.69	4.12×10^{-3}	1.44	9.334	49.31
$N(n\text{-}C_4H_9)_4^+$(BuFeFe)	8.66	14.78	1.80	9.402	53.88
$As(C_6H_5)_4^+$(AsFeFe)	8.30	3.62	1.73	9.445	57.40

From Bhattacharjee, A., Bhakat, D., Roy, M., Kusz, J. (2010). Electrical conduction property of molecular magnetic material—{N(n-C4H9)4[Fe(II)Fe(III)(C2O4)3]}∞: before and after thermal degradation. Physica B: Condensed Matter.

between the magnetization and conduction is made. Furthermore, the values of the activation energy do not give any indication of the mechanism by which the conduction occurs (Table 9.1).

BuFeFe is thermally degraded by being placed in a 750°C oven for 10 h. The product obtained shall be referred to as FeFe. The use of scanning electron microscopy shows that FeFe consists of tiny rod structures, whereas BuFeFe was powder-like. Conductivity of FeFe also follows the Arrhenius relation. The activation energy of FeFe is very similar to that of BuFeFe, although its conductivity is higher. It is proposed that the presence of hematite in FeFe is the cause.

The importance of this work lays in the exploration of the properties of new semiconductor molecular materials so that they can be best used. In addition, the synthesis of an oxide from the material via thermal degradation provides a new method for doing so.

SELF-ASSEMBLING NANOSTRUCTURES*

10

10.1 INTRODUCTION

10.1.1 NANOTECHNOLOGY: DEFINITION AND BACKGROUND

Nanotechnology is the term used to cover the control, design, and understanding of functional materials or structures on an atomic or molecular scale. The field of nanotechnology draws on principles from nanoscale science, engineering, and technology to further understand fundamental physical and chemical properties of materials and structures at the nanoscale. It has been proven that nanoscale matter does not necessarily exhibit the same properties observed at larger scales. In fact, much of the awe around nanoscale matter can be attributed to how physical, chemical, and biological properties, phenomena, and processes are substantially improved as compared to the same matter at the macroscale. The improved properties exhibited by materials at the nanoscale can be attributed to the distinctive structural features of a given material existing as an intermediate in size between isolated atoms and bulk macroscopic materials. In the range that defines the scale of nanotechnology between 1 and 100 nm, a material may exist in a different form of state or display different physical characteristics from those of the same material on a larger scale (Kelsall et al., 2005). Improved particle and material characteristics at the nanoscale will enable for the creation of advanced applications and may also lead to potential challenges.

Although nanotechnology as a subject exists primarily on a research basis, there continues to be a steady incline in the transition from a theoretical approach to practical applications of nanotechnology. However, before nanotechnology can fully revolutionize industry and technology, there needs to be greater fundamental scientific understanding and technological expertise to comprehend how these nanoscale particles or molecules function. More specifically, a better understanding of how nanomaterials organize and assemble themselves, how they are constructed, and how they operate in complex nanostructured systems needs is essential to develop the field of nanotechnology. This chapter will discuss the theory behind one approach to organize and assemble nanoscale particles or molecules: the self-assembly of nanostructured molecular materials and devices.

10.1.2 SELF-ASSEMBLY: DEFINITION AND BACKGROUND

Self-assembly of constituent molecules on the nanoscale order has existed in nature as soft materials assemble to produce cell membranes, biopolymer fibers, and viruses (Kelsall et al., 2005). Recently

---

*By Yaser Dahman, Gregory Caruso, Astrid Eleosida, and Syed Tabish Hasnain.

Nanotechnology and Functional Materials for Engineers. DOI: http://dx.doi.org/10.1016/B978-0-323-51256-5.00010-1

in laboratories, scientists and engineers have been able to fabricate nanoscale materials through a bottom–up approach known as self-assembly. Self-assembly is a process in which particles or materials at the nanoscale spontaneously arrange predefined components into ordered superstructures which can be exploited in various applications. Fabrication of nanostructures can take place in a stirred vessel either by static or dynamic self-assembly. Static self-assembly is more common between the two and involves systems at equilibrium conditions that do not dissipate energy. Formation of a structure by static self-assembly requires induced thermal effects on the system but the product is stable, whereas dynamic self-assembly takes place when the system dissipates energy (Kelsall et al., 2005). Further discussion behind the principles of self-assembly will be highlighted later in this chapter.

10.1.3 IMPORTANCE OF SELF-ASSEMBLY

Many scientists consider self-assembly to be the most promising and simple method to incorporate nanoparticles into functional structures (Yang, Gu, Li, Zheng, & Li, 2010). The ability to easily control self-assembly based on organic materials allows the manufacturer to specialize electronic, magnetic, or photonic properties of inorganic components to achieve advanced functionalities from materials that may be nonexistent, difficult to extract, or have a limited source in nature. Fabricated nanostructures possess advanced properties that may be applied usefully in various potential applications, and these advanced properties of nanostructures have attracted interest from various fields of industry including biomedicine, computers, electronics, robotics, telecommunications, transportation, and water treatment, just to name a few. In Section 10.5, articles have been reviewed to include specified applications for use in lithium-ion (Li-ion) batteries, supercapacitors, hydrogels, coagulants, optical filters, and visual displays.

Aside from the ability to formulate advanced properties of materials for specialized applications, the self-assembly fabrication method is considered to be the most feasible of the bottom–up approaches. Fabrication of self-assembled nanostructures in a stirred vessel is much more practical than other bottom–up nanostructure fabrication methods such as the arrangement of nanoparticles by hand or by a scanning tunneling microscope. In addition, various nanostructures of different morphologies can be produced through self-assembly by less effort (e.g., fewer hours of labor), lower cost (e.g., less material wasted and no need for expensive vacuum equipment), and in larger quantities than through top–down fabrication methods (Wang & Gates, 2009).

Although self-assembly has indicated promising results, industries are continuously investing into research and development to better understand the process of self-assembly to one day be able to implement it on a large-scale basis. Once self-assembly is fully understood, industries are hopeful to implement self-assembly for mass production of smaller, cheaper, faster, and overall better products. Complete understanding of nanotechnology and implementation of self-assembly and other fabrication methods may revolutionize the world as it is today.

10.2 BUILDING BLOCKS

Common applications of soft materials are found in soaps, plastics, and paints. The process of self-assembling materials at the nanoscale follows a bottom–up approach—molecules are capable of assembling themselves into given superstructures. In nanotechnology, soft matter can be manipulated by thermal stresses, thermal fluctuations, and chemical treatment to produce self-organizing soft materials such as liquids, polymers, and colloids to name a few. Self-organization of soft materials is also observed in nature through the development of cell membranes, biopolymer fibers, and viruses for example (Kelsall et al., 2005).

The self-assembly of amphiphiles, colloids, and polymers into mesophases influences the properties of soft materials and will be discussed as the building blocks in the formation of such materials. Soft materials can be divided into two categories: synthetic and biological. Synthetic, self-organizing soft materials are synthesized by mankind, whereas biological soft materials occur naturally and self-assemble by nature (Kelsall et al., 2005).

10.2.1 SYNTHETIC

The major building blocks of synthetic soft materials in current development at the nanoscale include the following: polymers, surfactants, lipids, colloids, and liquid crystals. Synthetic polymers can be made by a range of polymerization methods and techniques.

Conventional polymers spontaneously self-assemble into nanostructures (e.g., crystal lamellae in crystalline polymers). On the other hand, studies have been done in engineered self-assembly of polymers into designed structures, for example, microphase separation of block copolymers into nanostructures that become significant components in nanotechnology development (Kelsall et al., 2005).

Surfactants, which are surface-active agents, are made up of amphiphilic molecules containing hydrophilic tails and hydrophobic heads. On surfaces, the amphiphiles preferentially segregate and become active. An example of a surfactant is detergent. Another similar self-assembling building block is lipids which are biological amphiphiles. In water, lipids collectively gather together or aggregate into nanostructures to reduce the contact of the hydrophobic part of the amphiphiles with water molecules. Examples of nanostructures that are formed are micelles and vesicles. Micelles can be spherical or cylindrical and form hydrophobic cores and hydrophilic coronas in water. Vesicles are hollow spheres in which the outer surface is composed of layers of surfactant molecules (Kelsall et al., 2005).

Colloids are described as a microscopic substance dispersed evenly throughout another substance with dimensions in the range of 1 nm to 1 μm. Colloidal systems are heterogeneous in nature and are composed of two separate phases. Self-organization of molecules occurs when the dispersed phase comes into contact and mixes with a continuous medium. Common examples of colloids include aerosols, foams, and emulsions (Kelsall et al., 2005).

The final synthetic, self-organizing soft material that will be discussed is liquid crystals. Liquid crystals are composed of moderate size organic molecules with molecular order intermediate between that of a liquid and of a crystal. Liquid crystals can form into two different phases—thermotropic and lyotropic liquid crystal phases which are also termed as mesogens. Thermotropic phases are formed by organic molecules in the absence of solvent, whereas lyotropic phases are formed by amphiphiles in a solution. Fig. 10.1 shows common orientations of liquid crystal phases (Kelsall et al., 2005).

In the nematic phase, the positional order of the molecules is short range, oriented in an average direction as shown in Fig. 10.1A. In the smectic (layered) phase, the molecules are oriented in long-range translational order illustrated in Fig. 10.1B. This smectic phase has a high viscosity and is generally not useful for devices (Case Liquid Crystal Group, 2010). Self-assembly of liquid crystal phases is an important component in cell membranes and electronic displays (Kelsall et al., 2005).

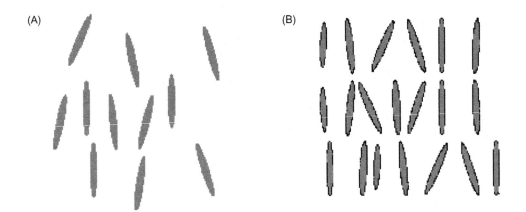

FIGURE 10.1

(A) Nematic phase and (B) smectic phase.

From Liquid Crystal Primer, Case Liquid Crystal and Complex Fluids Group, 2010.

10.2.2 BIOLOGICAL

Self-organization of soft materials also occurs naturally, as opposed to being synthesized, which is therefore convenient and advantageous. Examples of biological self-assembling materials are silk, collagen, proteins, DNA, microtubules, and viruses. Each will be discussed with current applications in nanotechnology developments.

The structure of silk consists of antiparallel β-sheets of protein fibroin, in which β-sheets are formed by intermolecular hydrogen bonding between peptide chains. The orientation of polypeptide chains along the β-sheets is responsible for the tensile strength of silk whereas weak forces in between the sheets allow for the flexibility of the material. Insects and arachnids produce silk to build webs, nests, and cocoons. Silkworm cocoons in particular have been used by mankind to produce high-quality fabrics (Kelsall et al., 2005).

Another self-organizing soft material is collagen which is made up of proteins that are naturally formed in living species. The structure of this material takes the form of cross-linked fibrils or may undergo denaturation by heat or chemical treatment to produce gelatine. It is a major component of connective tissues in animals. Similarly, keratin is also comprised of fibrous proteins that form hair, wool, nails, horns, and feathers on living species. In the natural form, keratin is arranged into fibrillar structures or by intermolecular hydrogen bonding of peptide chains (Kelsall et al., 2005).

Globular proteins are useful biological substances that make up enzymes, transport proteins, and may contain α-helix and/or β-helix secondary structures (Kelsall et al., 2005).

Relative to synthetic self-assembling materials, DNA fragments in a solution form lyotropic liquid crystals. In reference to Fig. 10.1, the short fragments behave like rods as illustrated, thereby forming liquid crystal phases. By increasing the concentration of the solution, self-organization of cholesteric (twisted nematic phase) and columnar liquid crystal phases may be observed (Kelsall et al., 2005). The structures of the preceding phases are shown in Fig. 10.2A and B.

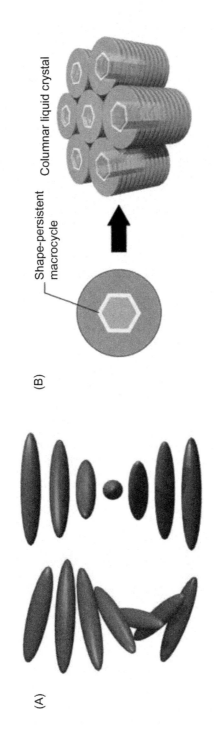

FIGURE 10.2

(A) Nematic twisting of cholesteric phase and (B) hexagonal columnar phase.

From (A) Barrett Research Group. (n.d.). Introduction to liquid crystals. Retrieved 31.03.11, from McGill University: http://barrett-group.mcgill.ca/teaching/liquid_crystal/ LC03.htm; (B) Organic and Biomolecular Chemistry. (2009). Retrieved 31.03.11, from Royal Society of Chemistry: http://www.rsc.org/publishing/journals/ob/article.asp?Ty pe=Issue&Journalcode=OB&Issue=16&SubYear=2009&Volume=7&Page=0&GA=on.

The biological self-organization of microtubules plays an important role in nanotechnology development. Microtubules are formed from protein tubulin. At the nanoscale, these structures can be used as nanochannels to transport liquids or as struts to support nanostructures. The importance of this substance in eukaryotes is of interest in the field of nanotechnology. Fig. 10.3 shows the structure of a microtuble (Kelsall et al., 2005).

Cilia, which are hair-like strands, move to allow fluid across the surface of organs, therefore the sliding of subfibers formed from microtubules shown in Fig. 10.3 influences the efficient motion of the cilia. In nanotechnology, it is currently under study to see if nanomotors like those used in cilia will be incorporated into nanomachines or will inspire designs for artificial motors (Kelsall et al., 2005).

The final biological self-assembling material that will be discussed is viruses. The structure of a virus takes the form of a sphere or helical (rod-like shape). Viruses are comprised of nucleic acid molecules encased in a protein coat and therefore often termed as virus capsids. Fig. 10.4 shows some of the common viruses (Kelsall et al., 2005).

Fig. 10.4A shows the self-assembled icosahedral structure of the herpes virus. The rhinovirus also takes on this structure. Fig. 10.4B and C illustrates the self-organized helical structure of the tobacco virus which was the first virus to be discovered. The dimensions of the tobacco mosaic virus in this figure are approximately 300 nm in length and 18 nm in diameter (Kelsall et al., 2005).

10.3 PRINCIPLES OF SELF-ASSEMBLY

Self-assembly is the reversible and cooperative assembly of predefined parts into an ordered structure, which assembles with no external influences after the initial trigger. Currently, self-assembly has been broken down into two categories: static and dynamic. Static self-assembly refers to systems at equilibrium which do not dissipate energy. The formation of the nanostructure may require energy, but the structure is stable once it has been formed. Dynamic self-assembly refers to the formation or patterning of structures when the system does, in fact, dissipate energy.

FIGURE 10.3

X-ray structure of a microtubule.

From Kelsall, R., Hamley, I., & Geoghegan, M. (2005). Nanoscale science and technology. Hoboken: John Wiley & Sons Ltd.

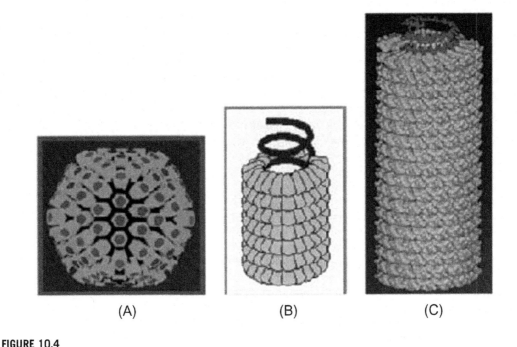

FIGURE 10.4

(A) Structure of herpes simplex virus protein shell; (B and C) tobacco mosaic virus.

From Kelsall, R., Hamley, I., & Geoghegan, M. (2005). Nanoscale science and technology. Hoboken: John Wiley & Sons Ltd.

Self-assembly in materials relies on the fact that the fluctuations in the orientation and position of the molecules or particles due to random movements have energies in the order of thermal energy. The use of thermal energy in changing material properties is quite prevalent in all industries, but the effect is even more important on the nanoscale. Thermal energy has a significant impact on materials on the nanoscale as non-covalent bonds are often broken and reformed in a new manner. Due to these non-covalent interactions between molecules, structure changes can be obtained by changes in the conditions provided for the molecules. For instance, temperature and pH changes help to initiate the transition of a structure to another.

10.3.1 NON-COVALENT INTERACTIONS

In order for self-assembly to occur, the non-covalent forces between the molecules need to be broken and reformed. As such, the forces between the molecules must be much weaker than the covalent bonds which hold the molecules together (Kelsall et al., 2005). In doing so, the molecules are not changed chemically, but are structured in a different orientation. The weak intermolecular interactions that govern molecular ordering in materials include hydrogen bonds, ionic interactions, dipolar interactions, van der Waals forces, and hydrophobic interactions.

Hydrogen bonding is especially important in biological systems. Protein structures in water are held together by hydrogen bonds (Kelsall et al., 2005). Hydrogen bonds are weaker than covalent bonds (about 20 kJ/mol compared to about 500 kJ/mol for hydrogen bonds and covalent bonds, respectively) (Kelsall et al., 2005). As a result, structures can self-assemble without chemical reactions needing to occur, and the bonds are strong enough to hold the structures together once they have been formed.

Dipolar interactions follow the same principles as hydrogen bonding, except they are not limited to just hydrogen atoms. Dipolar interactions refer to the direct interactions between two magnetic dipoles. The dipoles are a result of the difference in electronegativity within molecules creating partial positive and negative charges within the molecule.

The van der Waals forces are the sum of the attractive or repulsive forces between molecules—other than those due to covalent bonds. The forces include those between a permanent dipole and a corresponding dipole, as well as the London dispersion forces.

The hydrophobic effect arises when a nonpolar solute is inserted into water. The hydrophobic effect is attributed to the ordering of water molecules around a hydrophobic molecule. The ordering leads to a reduction in entropy (Kelsall et al., 2005). The entropy loss can be offset when association of hydrophobic molecules into micelles occurs, as this results in an increase in entropy.

10.3.2 INTERMOLECULAR PACKING

At higher concentrations, the packing of block copolymer or amphiphilic molecules in solution leads to the formation of lyotropic liquid crystal phases (Kelsall et al., 2005). These crystal phases include cubic-packed spherical micelles, hexagonal-packed cylindrical micelles, lamellae, and bicontinuous cubic phases. The lyotropic crystal phases are illustrated in Fig. 10.5.

The phase that forms is dependent on the curvature of the surfactant–water interface. To understand the lyotropic phase behavior, there exist two approaches. The first approach computes the free energy associated with curved interfaces; the curvature is analyzed using differential geometry, while not incorporating details of the organization of the molecules (Kelsall et al., 2005). The second approach uses a molecular packing parameter to describe the interfacial curvature (Kelsall et al., 2005).

10.4 METHODS TO PREPARE AND PATTERN NANOPARTICLES

10.4.1 NANOPARTICLES FROM MICELLAR AND VESICULAR POLYMERIZATION

One of the main challenges in nanotechnology is the fabrication of nanoparticles of certain, and controlled, size, shape, as well as functionality (Kelsall et al., 2005). There are several established routes to nanoparticle preparation. Spherical nanoparticles can be prepared by fine milling. Metal and metal oxide nanoparticles can be prepared using micellar nanoreactors.

Additionally, metal nanoparticles can be surface-patterned using the self-organization of block copolymers. For metal nanoparticles, a method of patterning is the nanoparticle formation occurring within micelles in solution, which is then deposited on a solid substrate. Another method of patterning metal nanoparticles is the direct patterning by way of selective wetting.

Nanocapsules can be prepared by employing cross-linking of the shell of the block copolymer vesicles (Kelsall et al., 2005). Another method is to use polyelectrolyte multilayers that are assembled around a colloidal core, which dissolves.

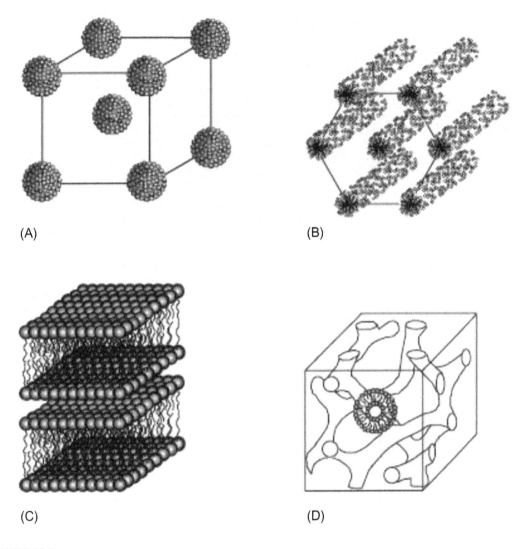

(A)

(B)

(C)

(D)

FIGURE 10.5

Lyotropic liquid crystal structures: (A) cubic-packed spherical micelles, (B) hexagonal-packed cylindrical micelles, (C) lamellar phase, and (D) bicontinuous cubic structure.

From Kelsall, R., Hamley, I., & Geoghegan, M. (2005). Nanoscale science and technology. Hoboken: John Wiley & Sons Ltd.

10.4.2 FUNCTIONALIZED NANOPARTICLES

With the ability to specify functionality for nanoparticles, there can be a whole new set of applications. Functionalized nanoparticles are already used in the medical field. Gold nanoparticles functionalized with proteins have been used as indicators to detect biological molecules, as well as changes in

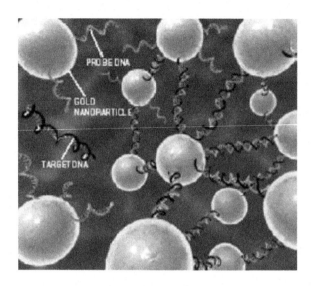

FIGURE 10.6

DNA-functionalized gold nanoparticle gene sequence detection system.

From Kelsall, R., Hamley, I., & Geoghegan, M. (2005). Nanoscale science and technology. Hoboken: John Wiley & Sons Ltd.

biological systems (Kelsall et al., 2005). Using dip pen nanolithography, arrays of nanoparticles can be prepared. A specific example is presented in Fig. 10.6, where functionalized nanoparticles are used for detecting specific gene sequences in genetic screening.

For detecting specific gene sequence, the sequence of bases in target DNA is identified. Two sets of gold nanoparticles are then prepared. One set has attached DNA that binds to one end of target DNA, while the other carries DNA that binds to the other end. Upon addition of target DNA, the two types of nanoparticles are bound together, forming an aggregate. The formation of the aggregate results in a shift in the light-scattering spectrum from the solution. A color change in the solution occurs and can be detected easily (Kelsall et al., 2005).

10.4.3 COLLOIDAL NANOPARTICLE CRYSTALS

Photonic band gap crystals have been receiving a lot of attention due to their ability to confine photons or control stimulated light emission, as well as constructing lossless waveguides (Kelsall et al., 2005). A photonic band gap crystal is a structure that has periodic variation in its dielectric properties. By forming face-centered cubic lattices from colloidal spheres, it has been learned that independent of the dielectric contrast, there is never a complete band gap. There is considerable effort being put forth into creating a complete three-dimensional (3D) band gap. In order to do so, the colloidal particles must have low dispersity (so as to form a cubic crystal), and the number of defects must be minimized (Kelsall et al., 2005).

Using so-called convective self-assembly, ordered crystals can be created by rapid evaporation of solvent, leaving the crystals alone on the surface. Another technique is the controlled withdrawal of the substrate from a colloidal solution, where lateral capillary forces induce crystallization of spheres; if the meniscus is swept slowly across the substrate, well-ordered crystals may be deposited (Kelsall et al., 2005). Finally, an epitaxial mechanism may be employed, using lithographically patterned polymer substrate to template crystal growth (Kelsall et al., 2005). For this mechanism, holes are created in a rectangular array, with sizes large enough to just hold one colloidal particle. Layer-by-layer growth can be achieved by slow sedimentation of the silica spheres that were used initially. Hollow or solid nanoparticles can be grown, allowing for the possibility of preparing colloidal crystals of solid or hollow TiO_2 particles, as well as conductive polymer nanoparticles (Kelsall et al., 2005).

10.4.4 SELF-ORGANIZING INORGANIC NANOPARTICLES

There is a growing interest in composite materials that have a polymer with filler particles such as clay particles. The fillers effectively modify the properties of polymer, improving transport, mechanical and thermal properties by 10-folds, and higher orders of magnitude. The polymer–clay nanoparticles tend to be lighter, while offering identical mechanical performance (Kelsall et al., 2005). Moreover, the mechanical properties for the polymer–clay nanoparticles are superior to fiber-reinforced and other polymers, because there is reinforcement in two dimensions from the inorganic layers, as opposed to one (Kelsall et al., 2005). Finally, the polymer–clay nanoparticles offer exceptional diffusional barrier properties, without requiring additional layers as part of the design. The particles are formed using silicate layers which are combined with layers of octahedrally coordinated atoms of aluminum or magnesium (Kelsall et al., 2005). The layers result in the clay being in a lamellar phase when in water.

10.4.5 LIQUID CRYSTAL NANODROPLETS

Liquid crystal droplets being patterned at the nanoscale has the potential for applications in phased array optics. In phased array optics, 3D images are reconstructed on a two-dimensional surface (Kelsall et al., 2005). Phased array optics allows this to happen if the amplitude and the phase of the light waves emitting from the virtual image are controlled. To reconstruct a light wave pattern, an array of switchable light sources 200 nm apart is sufficient in most cases (Kelsall et al., 2005). However, liquid crystals can be used as switchable phase shifters, hence their use in phased array optics. Patterning of liquid crystals in micelles is a proposed and promising method of arranging liquid crystals for nanoscale arrays (Kelsall et al., 2005).

10.4.6 BIONANOPARTICLES

Bionanoparticles are structures on the nanoscale that are biological, as opposed to synthetic. Viruses are natural nanoparticles, and a number of nanotech applications of viruses are now being considered. Viruses are used as responsive delivery agents. Modified cowpea chlorotic mottle virus nanoparticles have been used as biocompatible responsive delivery agents (Kelsall et al., 2005). The virus is a compact spherical structure, but as pH is increased above 6.5, the structure becomes porous, allowing the pH-controlled release of the drug. For nonviral gene delivery, synthetic polymers including polylysine and poly(oxyethylene)-based copolymers, biologically derived liposomes, cationic lipids, and cationic polyelectrolyte poly(ethyleneimine)—called PEI—have been studied. PEI is useful for binding anionic

DNA within the physiological pH range and forcing DNA to form condensates, which can be effectively transferred across the cell membrane (Kelsall et al., 2005).

10.4.7 NANOOBJECTS

Particles have been prepared with shapes other than the traditional spheres, shells, or tubes, at the nanoscale. Nanoprisms have been prepared using conversion via photoinduction. Nanocrystal growth can be controlled using organic agents, resulting in the production of polyhedra, with the growth rate of planes in crystal unit cell controlling the faces (Kelsall et al., 2005). String and other structures can be prepared using surfactants to selectively control the growth of certain crystal faces. Self-assembled nanostructures can also be employed to template the formation of helical nanoparticles, as well as vesicular or string structures formed by block copolymers (Kelsall et al., 2005).

10.5 CURRENT RESEARCH AND FUTURE PERSPECTIVES

Because self-assembly has much potential for a variety of applications, much research is dedicated to using and improving this method. This section discusses current research for applications in Li-ion batteries, supercapacitors, hydrogels, water treatment, and optical cable, displays, filters and sensors.

Hirst, Escuder, Miravet, and Smith (2008) reviewed the ability of self-assembled peptide nanofiber (hydrogel) to regenerate the optic tract and functional recovery of vision. Zhang and coworkers applied a nanoscaffold supported by a gelator with a peptide repeat unit of Arg-Ala-Asp-Ala to successfully enable reconnection of nerve tissue after optical tract in a hamster's midbrain had been surgically severed (Ellois-Behnke et al., 2006). The 1% concentrated peptide nanofiber solution was injected into the test hamster and regenerative results were compared to those of an untreated hamster. The optic tract in the treated hamster was able to regenerate and functional vision was restored, whereas no axonal regeneration in the untreated hamster was observed. From the same procedure, it was also discovered that the self-assembling peptide gel could be applied to a wound and achieve complete hemostasis in 15 s (Fig. 10.7).

Based on the results of Zhang, the authors propose that healing can be accelerated by close contact between the self-assembled nanofibers and the extracellular matrix to assist cell–scaffold interactions. Six months after hydrogel treatment, there was no support to indicate the existence of prion-like substances or fibril entanglements in the treated animals. The study concluded that there is potential to assemble hydrogels to generate or exhibit distinct forms of biological activity, at a time when there is a greater understanding of nanoscale self-assembly processes.

The field of electro-optic nanotechnology has many applications in sensors and other nanostructure devices. Tsang et al. (2008) discussed the self-assembly of nanocrystals into large colloidal crystals, in which the secondary structures are manipulated to fabricate nanoparticles as 3D regular structures, used for optical filters and other nanostructured devices. They developed a method to synthesize centimeter long magnetic colloidal crystals using magnetic gradient separation of chemical or biological entities. The large colloidal crystals are composed of 3D regular arrays of supermagnetic FePt nanoparticles encapsulated in a hydrophilic silica shell. The prepared sample underwent controlled crystallization followed by exposure to an external magnetic field. Fig. 10.8 shows the needle-shaped crystals approximately 1 cm in length and the stacking of nanoparticles.

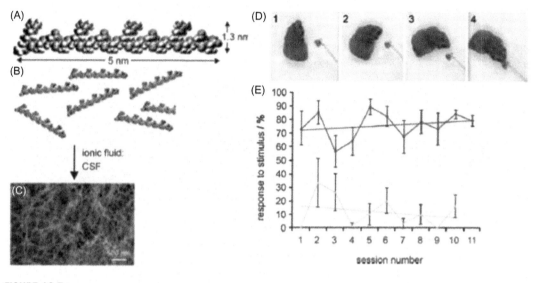

FIGURE 10.7

Peptide hydrogelators assemble into a fabrillar network as a consequence of hydrophobic and hydrogen bond interactions (A–C). When applied to the damaged optic nerve of hamsters blinded in their right eye, vision is regenerated and the hamster responds to stimulus (D). The data in (E) indicate that treated hamsters (blue (black in print version)) regained 80% vision, whereas untreated animals (yellow (gray in print version)) regained less than 20%.

From Hirst, A., Escuder, B., Miravet, J., & Smith, D. (2008). High-tech applications of self-assembling supramolecular nanostructured gel-phase materials: from regenerative medicine to electronic devices. Angewandte Chemie International Edition, 47, 8002–8018.

In this article, Tsang et al. (2008) shows the ability to manipulate secondary structured nanoparticles from self-assembly colloidal crystals into high-quality colloid crystals via magnetic crystallization of FePt-encapsulated silica shell. This approach opens up new development of high-quality nanostructured material that may potentially be used for optical filters, waveguides, sensors, and other nanodevices.

Link and Sailor (2003) discuss the development of Smart Dust—photonic crystals of porous Si—which is useful for various optics and sensor applications. They examine the method of electrochemical etching and surface chemistry, followed by the mechanism of self-assembly and self-orientation of colloidal photonic crystals to produce Smart Dust. The material is prepared using a two-step method. In the first step, an electric current is used in the electrochemical etching of Si, to produce a one-dimensional photonic crystal. In the second step, the porous Si photonic structure is chemically modified. As a result a thin film is produced and fractured into nanostructured smaller particles with the ability to self-assemble. Fig. 10.9 is a detailed flow diagram of the synthesis of Smart Dust particles.

The preceding technique manipulates matter at the nanoscale to develop nanostructured material and devices by self-assembly. In this study, when introduced at a liquid interface, the fractured Smart Dust particles self-orient spontaneously with the hydrophobic side facing up. Smart Dust particles can be

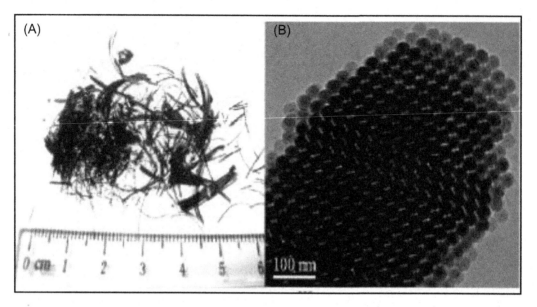

FIGURE 10.8

(A) Needle-shaped crystals—3 nm FePt colloidal crystals inside 33 nm silica sphere and (B) transmission electron microscopy (TEM) image of hcp arrangement of encapsulated nanoparticles.

From Tsang, S., Yu, C., Tang, H., He, H., Castelleto, V., Hamley, I., et al. (2008). Assembly of centimeter long silica coated FePt colloid crystals with tailored interstices by magnetic crystallization. Chemistry of Materials, 4554–4556.

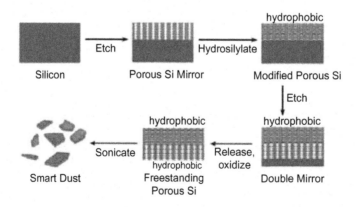

FIGURE 10.9

Synthesis of bifunctional Smart Dust particles.

From Link, J. R., & Sailor, M. J. (2003). Smart Dust: self-assembling, self-orienting photonic crystals of porous Si. Proceedings of the National Academy of Sciences, 10607–10610.

used in information display, optical switching, chemical and biological sensor fields, high-throughput screening, and drug discovery.

Zhang, Le, Malalur-Nagaraja-Rao, Hsu, and Chiao (2005) propose a solution to challenges of manufacturing in the optical communication and sensor industry. Micro-optical components require expensive alignment and assembly procedures. The development of a self-assembled, micro-optical component may thus be a solution. Self-assembly of 3D microstructures was designed to eliminate manual or robotic manipulations that are costly and sometimes inaccurate. The design was conducted using MEMS fabrication foundry service—MUMPS (Multi-user MEMS Process). Fig. 10.10 shows the mechanical design of the device. A scratch disk actuator is hinged to a polysilicon frame. The direction of motion of the actuator is a straight path controlled by applied voltages. The hinges allow for out-of-plane, perpendicular assembly.

The application of a polymer on the polysilicon frame results in good optical transparency and shows good mechanical properties for the assembly process. Optical performance of the filter had been yet to be examined.

Certain optical and electronic devices are fullerene-based. Georgakilas, Pellarini, Prato, Guldi, and Melle-Franco (2002) examined an assembly method to produce the formation of nanotubular structures. The authors decided on using fullerenes for their self-assembly process because they have excellent electronic properties and because their derivatives were found to self-assemble at the nanoscale. Four different compounds were prepared and immersed in ultrasonic baths before being transferred to TEM. Fig. 10.11A and B shows TEM images for the first and second compounds. It was observed

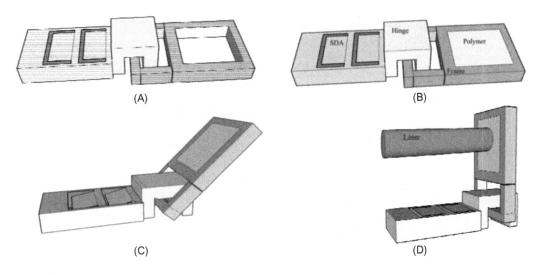

FIGURE 10.10

Device architecture. (A) Scratch drive actuators (SDA) is connected to hinged polysilicon frame; (B) polymer film acts as optical interface on polysilicon frame; (C) actuators lift up frame; (D) the frame stands perpendicular to the substrate after assembly with a laser shone normal to the filter surface.

From Zhang, P., Le, K., Malalur-Nagaraja-Rao, S., Hsu, L., & Chiao, J. (2005). Self-assembly micro optical filters. Society of Photographic Instrumentation Engineers.

that spheres were formed by compound 1, while bundles of nanorods were formed by compound 2. Fig. 10.11C shows nanorods formed by compound 3, while compound 4 produced neither rods nor spheres, but isolated tubules. They also observed that ligation of compound 2 with a porphyrin moiety promoted the formation of nanorods with strongly improved shaping.

Much nanotechnology research is continuously performed in the electrical and electrochemical fields. One important application is making and improving energy storage devices. Xiong, Yuan, Zhang, and Qian (2011) examined the potential advantage of NiO nanostructures as supercapacitor materials. They used NiO quasi-tubular structures by self-assembly of nanoparticles as basis for illustration (Fig. 10.12). They prepared various NiO hierarchical nanostructures by heating the corresponding self-assembled NiO hierarchical nanostructure precursors at 300°C for 2 h at a heating rate of 1°C/min. The self-assembled NiO quasi-tubular nanostructure in an amount of 75 wt% was mixed with

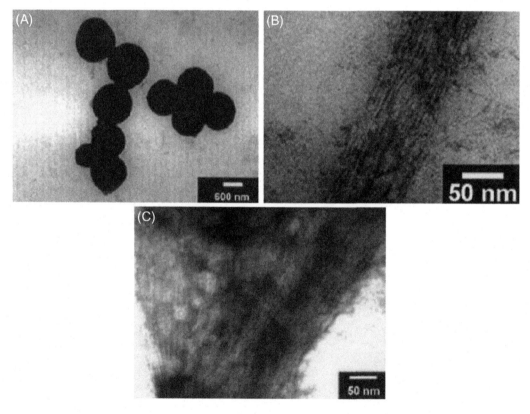

FIGURE 10.11

(A) TEM images for compound 1. The images showed spheres being formed. (B) TEM image of a bundle of nanorods formed by compound 2. (C) TEM image of a bundle of nanorods formed by compound 3.

From Georgakilas, V., Pellarini, F., Prato, M., Guldi, D. M., & Melle-Franco, M. (2002). Supramolecular self-assembled fullerene nanostructures. Proceedings of the National Academy of Sciences, 5075–5080.

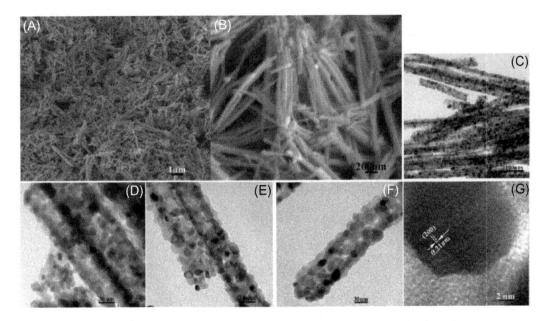

FIGURE 10.12

(A and B) Field-emission scanning electron microscopy (SEM) images and (C–F) TEM images of the as-obtained NiO quasi-tubular structures by self-assembly of nanoparticles. (G) High-resolution TEM image taken from an individual nanoparticles self-assembling into NiO quasi-tubular structures.

From Ziong, S., Yuan, C., Zhang, X., & Qian, Y. (2011). Mesoporous NiO with various hierarchical nanostructures by quasi-nanotubes/ nanowires/nanorods self-assembly: controllable preparation and application in supercapacitors. CrsytEngComm, 13, 626–632.

20 wt% of acetylene black and 5 wt% of poly(tetrafluoroethylene) and pressed (20 MPa) onto a nickel grid (1 cm^2) for assembly of the test electrode. These rich in porosity nanostructures have large specific surface areas that promote efficient contact with more electroactive sites for electrolyte, even at high current densities.

They observed the self-assembled quasi-tubular structures to display high specific capacitance (SC) and its ability to retain SC at higher current densities. The NiO quasi-nanotubes are able to maintain a SC of 345 F/g after 1500 continuous charge–discharge cycles, relative to the initial SC of 314 F/g. Therefore, the NiO quasi-tubular structures self-assembled with nanoparticles have demonstrated their excellent capacity retention of 91% after 1500 continuous charge–discharge cycles and high rate capability. Excellent electrochemical performance data of these easy to assemble NiO quasi-tubular structures suggest they have potential use in applications such as supercapacitors and Li-ion batteries.

Yang et al. (2010) examined the ability of a self-assembled 3D flower-shaped SnO$_2$ nanostructure to be used as an anode in Li-ion batteries. The flower-shaped SnO$_2$ nanostructure was self-assembled by hydrothermal treatment of a mixture of tin (II) dichloride dehydrate (SnCl$_2$·2H$_2$O) and sodium citrate (Na$_3$C$_6$H$_5$O$_7$·2H$_2$O) in a continuously stirred solution of distilled water and sodium hydroxide (NaOH) (Yang et al., 2010). The electrochemical performance of the SnO$_2$ nanostructure in the potential range of 0.005–1.5 V was tested for up to 30 cycles and the results were compared against commercially used carbon-based materials. The authors observed that nonconductive impurities (such as hydroxides and organic material) that remain on the nanostructure following the hydrothermal synthesis reduce the overall conductivity and the irreversibility during charge and discharge cycles. To solve the issue, the SnO_2 nanostructures were heat-treated in air above 400°C to remove the hydroxides and decompose the organic impurities. The electrochemical performance of the heat-treated nanostructures was improved over the unheated sample. Fig. 10.13 depicts SEM images of the SnO_2 heat-treated at various temperatures, and Fig. 10.14 compares the electrochemical properties of the heat-treated samples from Fig. 10.13.

When the flower-like SnO_2 nanostructures are used as anode material for Li-ion batteries, they demonstrate a reversible capacity of approximately $670\,mA\cdot h\cdot g^{-1}$ after 30 cycles, which is almost double

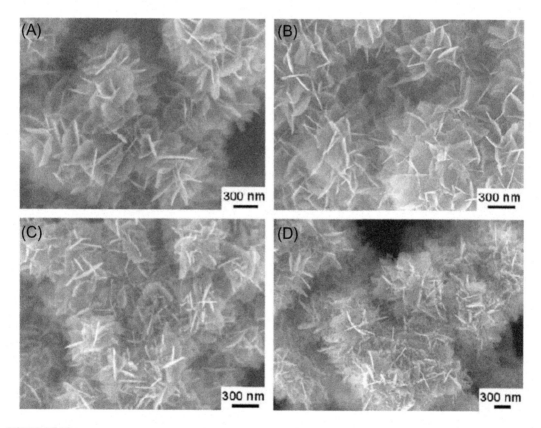

FIGURE 10.13

SEM images of the SnO_2 nanostructures prepared after calcinations in air at different temperatures: (A) as-synthesized nanostructures obtained by hydrothermal treatments; (B) after heating in air at 300°C, (C) 400°C, and (D) 700°C, respectively.

From Yang, R., Gu, Y., Li, Y., Zheng, J., & Li, X. (2010). Self-assembled 3D flower-shaped SnO₂ nanostructures with improved electrochemical performance for lithium storage. Acta Materialia, 58, 866–874.

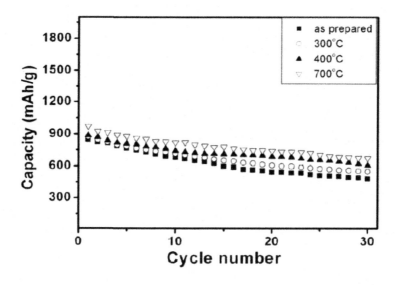

FIGURE 10.14

Charge capacities of the heat-treated (in air) SnO$_2$ nanostructures.

From Yang, R., Gu, Y., Li, Y., Zheng, J., & Li, X. (2010). Self-assembled 3D flower-shaped SnO$_2$ nanostructures with improved electrochemical performance for lithium storage. Acta Materialia, 58, 866–874.

the reversible capacity of the commercially used carbon-based materials ($372\,mA{\cdot}h{\cdot}g^{-1}$) (Dahn, Zheng, Liu, & Zue, 1995). In addition, the average capacity loss of the flower-like 3D nanostructure was estimated to be 0.95% per cycle af ter the second cycle, smaller than that of pristine SnO$_2$ particles (3.46%). The study concluded that the improved electrochemical performance of the self-assembled 3D SnO$_2$ nanostructures indicates that these nanostructures can enhance the electrochemical properties of anode materials for Li-ion batteries.

There are many other studies that may prove useful for Li-ion battery improvement. Wang et al. (2009) examined the insertion and extraction kinetics for Li ions using functionalized graphene sheets (FGSs). They created hybrid nanostructures to observe the effects on the kinetics for Li ions during insertion and extraction. TiO$_2$ was chosen as the metal oxide and prepared as the hybrid with FGS. Both rutile and anatase TiO$_2$ were used to study the effects. Using sonification, sodium dodecyl sulfate (SDS)–FGS dispersions were prepared. The SDS surfactant helps promote the formation of the TiO$_2$ hybrid nanostructures, as it determines the interfacial interactions between graphene and the oxide materials. As crystalline TiO$_2$ formed, nanoparticles began coating the graphene surface due to the sulfate head groups having strong bonding with TiO$_2$. The hybrid nanostructure, TiO$_2$–graphene, enhanced Li-ion insertion and extraction kinetics, particularly for higher charge and discharge rates. Fig. 10.15A and B shows the improved specific capacity for the hybrid nanostructures over the metal oxides alone.

From this, it was concluded FGS should serve as a conductive additive for Li-ion battery electrode materials to improve diffusion, electron transport, and lower the resistance at the interface of electrodes and electrolytes.

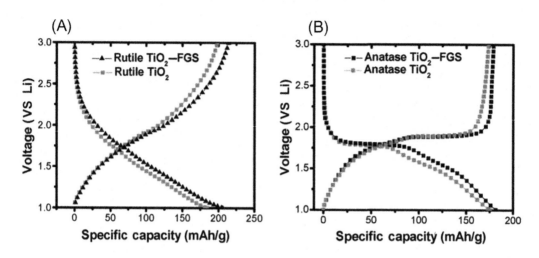

FIGURE 10.15

(A) Charge–discharge profiles for control rutile TiO_2 and rutile TiO_2–FGS hybrid. (B) Charge–discharge profiles for anatase TiO_2 and anatase TiO_2–FGS hybrid.

From Wang, D., Choi, D., Li, J., Yang, Z., Nie, Z., Kou, R., et al. (2009). Self-assembled TiO_2-graphene hybrid nanostructures for enhanced Li-ion insertion. American Chemical Society, 907–914.

Self-assembly has also been studied outside the electronic and electro-optical fields. For example, Zhong, Hu, Liang, Cao, and Song (2006) investigated synthesizing an iron oxide precursor with 3D flower-like nanostructure using an ethylene glycol-mediated process, and then using it for treating wastewater. Ferric chloride was used as the source for iron. Ferric chloride, urea, and tertbutylammonium bromide were dissolved in ethylene glycol. Once the iron oxide precursor had been formed, it was collected and the morphology of the precursor was studied by SEM. Fig. 10.16 shows the formation of the 3D flowerlike nanostructures as the reaction proceeds. Fig. 10.16A shows the nanoparticles being formed, then beginning to cluster in (B). In Fig. 10.16C, the clusters start taking the flowerlike shape, and by (D) the sample became composed of entirely 3D flowerlike nanostructures.

Once the nanostructures were created, they were employed in wastewater treatment. Arsenic and chromium were the two major components that needed to be removed, and the iron oxide was able to do so with an adsorption capacity of 5.3 mg As/g and 4.47 mg Cr/g. It was concluded that the as-prepared iron nanostructures should be able to remove heavy metal ions and other pollutants in water treatment extremely well.

10.6 CONCLUSIONS

In this chapter, the definition and importance of self-assembled nanostructures were highlighted. The principles of self-assembly were discussed and applications of self-assembled nanostructures were reviewed. Self-assembly is considered to be one of the simplest methods to synthesize nanostructures.

FIGURE 10.16

SEM image as the reaction proceeded at time (A) 2 min, (B) 6 min, (C) 7 min, and (D) 8 min.

From Zhong, L., Hu, J., Liang, H., Cao, A., & Song, W. (2006). Self-assembled 3D flowerlike iron oxide nanostructures and their application in water treatment. Advanced Materials, 2426–2431.

However, there still remains a great challenge to fabricate hierarchically self-assembled nanostructures with specified chemical components and controlled morphologies (Yang et al., 2010). Nanostructures can be self-assembled from molecular materials and devices to exhibit advanced properties that can be achieved through the application of external fields, typically by simple hydrothermal or magnetic treatment.

The primary focus of Section 10.5 is the fabrication of 3D nanostructures prepared by simple hydro-thermal or magnetic self-assembly processes. The formulated 3D nanosuperstructures were applied as components in Li-ion batteries, supercapacitors, hydrogels, water treatment coagulants, and optical filters and sensors. Researchers have either proven advanced physical and chemical properties of the self-assembled nanostructures in specified applications or they have yet to explore that in research and development. Potential applications of the assembled nanostructures are suggested to provide insight into the broad spectrum to which their respective features can be applied.

Significant progress has been reported in the development and use of self-assembled nanostructures for application of their advanced properties, including the preparation of nanoparticles, patterning of nanostructures, nanomotors, utilization of biomineralization, and for functionalized drug delivery (Kelsall et al., 2005). It is likely that the first nanostructures to be commercialized will be nanoparticles fabricated in micellar or vesicular nanoreactors, and catalysts and separation media fabricated from mesoporous templated materials (Kelsall et al., 2005). Self-assembly methods have proven to fabricate effective, efficient, and low-cost nanostructures. In the next few decades, there will be a creation of many new materials and nanosuperstructures as novel research in the field of nanotechnology is expected to increase exponentially. As the case with most scientific and technological breakthroughs, the rise and development of nanotechnology and self-assembled nanostructures can be expected to impact the quality of life of people everywhere—for better or for worse.

NANOMEDICINE*

11.1 INTRODUCTION TO NANOMEDICINE

Nanomedicine is the application of nanotechnology to medicine and is based on three mutually overlapping and progressively more powerful molecular technologies, which are listed below (Jain, 2008):

1. Nanoscale-structured materials and devices
2. Molecular medicine via genomics, proteomics, and artificially engineered microorganisms
3. Molecular machine systems.

Nanomedicine is different from conventional approaches in a sense that conventional medicine's approach is based on a response to problems at tissue level and by the time that symptoms become apparent to the patient or clinician, the disease may often be at an advanced stage. Conventional technology also relies more on the use of sophisticated and expensive medical tools, machines, robots, minimally invasive devices, and implants. However, in nanomedicine, the goal is to detect changes and problems at the molecular and cellular levels and to treat them before they spread out (Moore, 2009).

Since nanotechnology is able to improve prognosis and focuses more on nanoscale interactions within individual cells, nanomedicine can be engineered to help the body to repair or replace lost or damaged tissue rather than to destroy and replace it with nonbiological material. Currently it is possible to a limited number of comparatively simple human tissues but, ultimately, the goal of regenerative medicine will be to replace more complex tissue systems such as bone, blood vessels, nerves, or to replace or partially replace damaged organs which are generally comprised of a number of different specialized cell and tissue types (Moore, 2009).

Nanotechnology is also being applied to diagnostic testing and monitoring. Many diagnostic devices rely on small amounts of analytes, such as blood and other body fluids being sampled, separated, diluted or mixed, and then being brought into contact with other biomolecules, sensors, or measuring systems. With nanomedicine, it is possible that such functions are being integrated using a highly multidisciplinary approach. Lab-on-a-chip systems are increasingly employing nanoscale features to the extent that only a few picoliters of analyte may be required (Moore, 2009).

Besides regenerative medicine and diagnostic testing and monitoring, nanotechnology is also being harnessed to use the power of less-invasive imaging techniques, like MRI. Here, magnetic or paramagnetic nanoparticles, often functionalized with the use of targeting biomolecules that bind to specific cell receptors, can be used in conjunction with MRI to track the movement of the injected nanoparticles to

---

* By Yaser Dahman, Thai Truong, Mahrukh Jawaid, and Sherlly Widjaja.

Nanotechnology and Functional Materials for Engineers. DOI: http://dx.doi.org/10.1016/B978-0-323-51256-5.00011-3

the desired target site. When accumulated at the target site, the magnetic fields produced by the MRI device can be used to heat the nanoparticles by a few degrees which are sufficient to kill the cancer cells locally without damage to surrounding tissues (Moore, 2009).

11.2 APPLICATIONS OF NANOTECHNOLOGY IN NANOMEDICINE

Nanotechnology is the creation and utilization of materials, devices, and systems through the control of matter on the nanometer. Nanotechnology already has an impact on health care. Several advancements have been made in almost every field of medicine. The accomplishments by the scientific community are tremendous and the future prospects are wide open. A number of fields in which the role of nanotechnology is most prominent are listed below (Jain, 2008):

- Nano-ophthalmology
- Nanooncology
- Nanoneurology
- Nanocardiology
- Nanoorthopedics
- Nanopharmaceuticals
- Nanotechnology in Biological Therapies
- Nanodevices in Medicine and Surgery
- Regenerative Medicine and Tissue Engineering
- Nanopulmonology
- Nanodermatology.

In this chapter, we will explain how nanomedicine is applicable to bone implant, tissue engineering, ophthalmology, and pharmaceuticals.

11.3 NANOPHARMACEUTICALS

Nanopharmaceuticals consist of discovery, development, and delivery of drugs. Discovery is the screening of drugs and finding which drugs work best. The discovery process needs to be improved and one way to improve the process is by using nanotechnology to create massive arrays of small spaces. This would allow for direct readings of signals using microfluidics. The idea of minimalizing lab space is comparable to computers that used to take up an entire room, while now they can fit in the palm of the hand. To be more efficient, researchers are aiming to turn lab experiments that can fit on the nanoscale for drug screening, so they can do thousands at a time. Some popular nanoparticles that are used for drug screening are gold nanoparticles, quantum dots, nanolasers, etc., which will be discussed here (Jain, 2008).

11.3.1 DRUG DISCOVERY

Gold nanoparticles do not burn out or blink compared to other nanoparticles like fluorophores or semiconductor nanoparticles. Other advantages are that they are easy to prepare, they have low toxicity, and they can easily be attached to the biological molecule of interest. Furthermore, the laser used to visualize these particles is of a frequency that does minimal damage to biological tissues. This technology is used for single-molecule tracking for drugs and biological samples (Jain, 2008).

Quantum dots are also used for single-molecule tracking. Compared to fluorescent dyes, quantum dots are brighter, more stable, and smaller, which make them more ideal for tracking. They also last much longer than dyes of up to 40 min compared to 5 s. Quantum dots are used to observe processes that last several minutes. One example is observing receptors to develop a drug to fight epilepsy and depression (Jain, 2008).

Nanolasers can help find drugs to halt the progression of Alzheimer's disease (AD) or Parkinson's disease (PD), and illnesses caused by radiation or chemical nerve agents. In AD and PD, the mitochondrion is one of the cell's organelle that is affected. Finding drugs to protect the mitochondrion has been difficult and slow. The process can be sped up by using nanolasers. Researchers have used lasers similar to those used in compact disc and combined them with living cells to detect death throes of mitochondria. Another technique with lasers was by flowing mitochondria through a solid-state microscopic laser and found that the mitochondria will begin "lasing" themselves. The frequency given off by the mitochondria was a function of their health. Healthy mitochondria gave off one color and swollen or dying mitochondria gave off another. The technology could be used by giving the mitochondria the death signal that it gets from AD or PD. Then thousands of drugs can be tested simultaneously that can inhibit the death signal. Currently the most popular drug used is cyclosporine but this drug weakens the immune system. Researchers hope to find a better drug using this laser technology (Jain, 2008).

Atomic force microscopy (AFM) is normally used for imaging at the nanoscale, but it can be used for drug discovery. A ligand can be attached to the tip of an AFM and the functionalized tip can be used to probe biological systems. One can modify living conditions of the cells and see how they react to the functionalized tip. By doing this, we can explore many different biological systems and learn their chemical entities and properties on the surface. Using AFM and functionalizing the tip has been used in AD. A peptide called Aβ is involved in the mechanism of AD, and the problem that the peptide has is it converts from being soluble to insoluble. AFM was used to see which antibodies would be able to inhibit the morphology of the peptide. Two antibodies were tested (M266.2 and m3D6). M266.2 completely inhibited the morphology and m3D6 only slowed the morphology of the peptide. This was one example AFM can be used to learn chemical properties of a biological system (Jain, 2008).

A device was created that can handle thousands of experiments simultaneously. The device was about 1 cm^2 that held thousands or even millions of vessels. The bottom was sealed and the top was open, which resembled nano test tubes. They are able to use this chip to screen for drugs millions at a time. One suggestion is for chemotherapy. The membranes of cancer cells normally have pumps that pump out the chemotherapy drug. This chip can be used to find a drug that inhibits the pumps, allowing for a more efficient chemotherapy drug (Jain, 2008).

11.3.2 DRUG DEVELOPMENT

Dendrimers are core–shell shaped, synthesized for many applications. Dendrimers are normally used for drug delivery, but they can be made as drugs as well. Specialized chemistry allows for precise control over the chemical and physical properties of dendrimers. Multiple drugs can be attached to dendrimers to be used for cancer therapeutics. The advantages to using dendrimers as drugs are (Jain, 2008):

- Tailor-made surface chemistry
- Nonimmunogenic
- Inherent body distribution enabling appropriate tissue targeting
- Possibly biodegradable.

Nanobodies are antibodies lacking light chains and are fully functional. Like conventional antibodies, they have high target specificity, low toxicity, but they are also like small drugs and are able to inhibit enzymes and access receptor clefts. Their advantages are (Jain, 2008):

- They combine the advantages of antibodies and small drugs.
- They can reach therapeutic targets that conventional antibodies cannot, for example, active sites of enzymes.
- They are stable.
- They can be administered by any means other than injections.
- They are cheap to produce on a large scale.

The selection and cloning of antibodies eliminate the need to construct and screen large libraries, and in vitro affinity maturation steps. Nanobodies' unique properties allow them to excel compared to conventional antibodies because they recognize uncommon or hidden epitopes, bind into cavities or active sites of protein targets, tailoring of half-life, drug format flexibility, low immunogenic potential, and ease. These properties make nanobodies a good potential for therapeutic agents (Jain, 2008).

11.3.3 DRUG DELIVERY

Dendrimers are normally used for drug delivery because of their physical properties and well-defined molecular weight. Their advantages for drug delivery are (Jain, 2008):

- They are a highly branched polymer, so several drugs can be attached.
- When they are administered into the human body, the pathway can be reproducible because of their well-defined molecular weight.
- Their biological properties can be changed to discover new effects related to their new architecture.

Researchers at the University of Michigan found a fast way to make multifunctional dendrimers using DNA to bind the dendrimers together. They make dendrimers that have one attachment, say a tracker molecule, a drug, or a ligand. Then attach what they called ssDNA strands to each dendrimers. These dendrimers link naturally by the DNA strands. This will allow a library full of dendrimers and they can link and make combinations of their choosing (Jain, 2008).

Micelles size range from 50 to 200 nm. The inside of a micelle is hydrophobic and the outside is hydrophilic. Micelles are useful because insoluble drugs could be put inside the hydrophobic part and the micelle itself is hydrophobic, making the drug essentially soluble. One way to release the payload is by making the micelles pH-sensitive. When the micelle reaches its target site and the pH of the cell is known. The pH-sensitive micelle will dissolve at the pH of the cell and release the payload (Jain, 2008).

Nanotubes are cylindrical tubes at the nanoscale. They are very strong and offer protection to their payload during delivery. The payload is put inside the cylinder and the outside is usually coated with a charge, ligands, or other functional nanoparticle to make the nanotube target specific. The ends are sometimes capped to keep the payload inside and can be removed by a biological signal when it reaches its target (Jain, 2008).

Liposomes properties vary with lipid composition, size, and surface charge, so they are categorized into three different classes (Jain, 2008).

1. Small with a single lipid layer
2. Larger ones with single lipid layer
3. Multi-lipid layer and in between the layers are aqueous solutions.

Liposomes are not great for drug delivery compared with the other methods mentioned. They are instable, unable to reach the proper sites, and unable to release the payload upon arrival. However, they can be coated to become more specific (Jain, 2008).

11.3.4 RESEARCH FINDINGS IN NANOPHARMACEUTICALS

11.3.4.1 The influence the properties of nanotubes can have on the cytotoxicity of cancer cells

Silicon nanotubes (SNTs) are a new class of inorganic nanotubes that can be used for bioseparation, drug delivery, imaging, and other biomedical applications. The use of SNTs to be used in biomedical applications is still in its early stages of research. Some properties that need to be further investigated are structural and functional properties that influence the biocompatibility and cellular uptake of the nanotubes. Some of these structural and functional properties are discussed below (Nan, Bai, Son, Lee, & Ghandehari, 2008).

In an experiment done by Nan et al. (2008), they synthesized SNTs using "template synthesis" combined with surface sol–gel chemistry (Fig. 11.1A). Two different lengths were made, namely 200 nm and 500 nm with 50 nm diameters. The effect of charge was also of interest, so positively charged tubes and non-charged tubes were made. Fig. 11.1D and E shows the tubes of discrete monodispersed

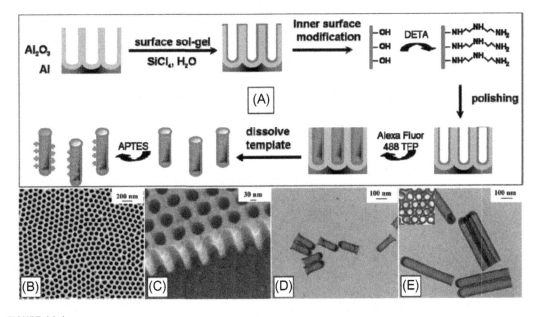

FIGURE 11.1

(A) Synthesis of silica nanotubes using a template. (B and C) Scanning electron microscopy (SEM) images of the templates. (D and E) Transmission electron microscopy (TEM) images of nanotubes.

From Nan, A., Bai, X., Son, S. J., Lee, S. B., & Ghandehari, H. (2008). Cellular uptake and cytotoxicity of silica nanotubes. Nano Letters, 2150–2154.

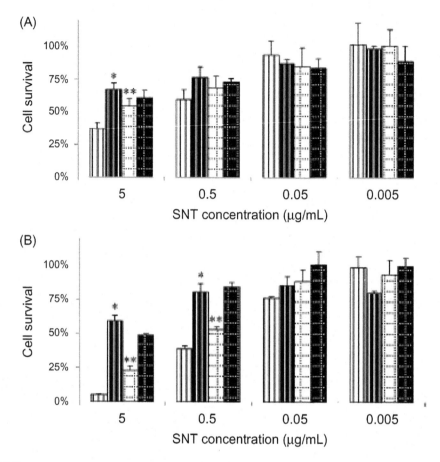

FIGURE 11.2

Cytotoxicity of varying concentrations of nanotubes and cell survival for MDA-MB-231 (A) and human umbilical vein endothelial cell (HUVEC) (B). *Results are reported as mean ± standard error of mean (n = 3). Statistically significant differences are indicated relative to 200 nm positively charged SNTs (*, p < 0.002; **, p < 0.02).*

From Nan, A., Bai, X., Son, S. J., Lee, S. B., & Ghandehari, H. (2008). Cellular uptake and cytotoxicity of silica nanotubes. Nano Letters, 2150–2154.

dimensions. The length and diameters can be controlled by changing the templates pore diameters and length to create nanotubes of different dimensions (Nan et al., 2008).

The effect of size and charge of the SNTs cytotoxicity to cancer cells was evaluated against an epithelial breast cancer cell line (MDA-MB-231) and a primary HUVEC. The cell survival of the cancer cells was concentration dependent. At concentrations of 0.05 µg/mL and lower, the SNTs had little or no effect on the cancer cells. At higher concentrations of 0.5 and 5 µg/mL both sizes of SNTs were able to inhibit breast cancer cells growth by up to 63%. In terms of the charge on the SNTs, the positively charged SNTs were significantly more toxic (Fig. 11.2A). The finding might suggest that the positively charged SNTs are more attracted to the negatively charged membrane of the breast cancer cells (Nan et al., 2008).

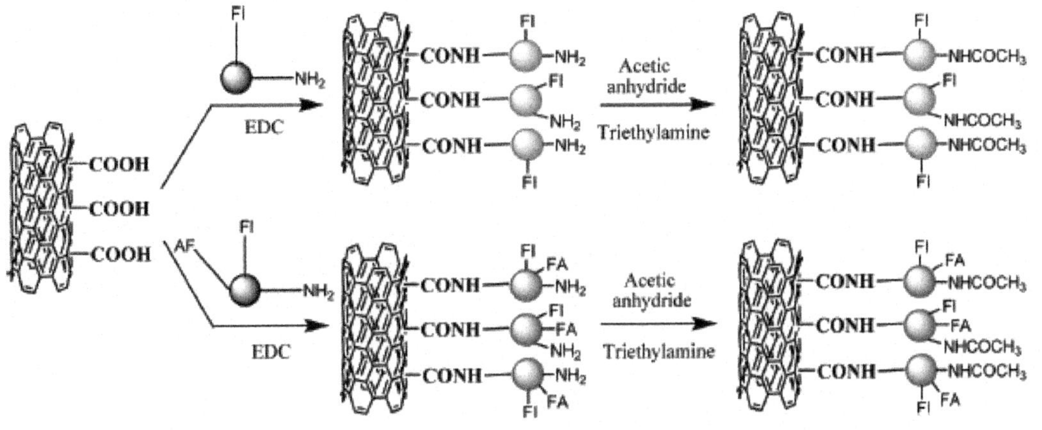

FIGURE 11.3

Process schematic of how the MWCNT was functionalized.

From Shi, X., Wang, S. H., Shen, M., Antwerp, M. E., Chen, X., Li, C., et al. (2009). Multifunctional dendrimer-modified multiwalled carbon nanotubes: synthesis, characterization, and in vitro cancer cell targeting and imaging. Biomacromolecules, 1744–1750.

The SNTs showed similar concentration trends for the HUVECs (Fig. 11.2B). However, at high concentrations, the effect of size was much more toxic in the HUVECs than the MDA-MB-231. The positively charged SNTs showed up to 10-fold more toxicity in the HUVECs than the MDA-MB-231. These results indicate that concentration and surface charge are cell-type dependent.

At concentrations of 0.5 and 5 μg/mL, the smaller length SNTs were more toxic for both cancer cells, indicating that toxicity is size dependent. In a literature by Medina et al. (2007) discussed that smaller sized particles generally interact more than larger particles and would allow for enhanced intrinsic toxicity. Another explanation is that for a given concentration the amount of 200 nm SNTs are greater than 500 nm SNTs by about 2.5 given they have the same density. Therefore, there number of SNTs are greater and would have a greater chance of interaction with cancer cells for the 200 nm SNTs (Nan et al., 2008).

11.3.4.2 Multifunctional carbon nanotubes and their effect on breast cancer cells

In a study done by Shi et al. (2009), they functionalized a multi-walled carbon nanotube (MWCNT) using G5 dendrimers that were covalently bonded to fluorescein isothiocyanate (FI) and folic acid (FA). To attach the dendrimers, the carboxyl residues on the surface of the MWCNT were treated with strong acid so the dendrimers could conjugate with the amine groups of the dendrimer. The remaining amine groups were acetylated to neutralize the MWCNT and resulted in a MWCNT/G5·NHAc-FI-FA complex. A control complex was made without the FA (MWCNT/G5·NHAc-FI). The reaction process of how the MWCNT is made, is shown in Fig. 11.3.

A series of experiments were done to ensure the proper functionalized MWCNT was produced:

1. UV–vis spectra were used to see if the MWCNTs were functionalized. With no functionalization, a 255 nm peak is displayed which is normal for MWCNT materials. The functionalized MWCNTs showed an absorption peak at 500 nm which is the peak that characterizes the attachment of the dendrimer (Shi et al., 2009).

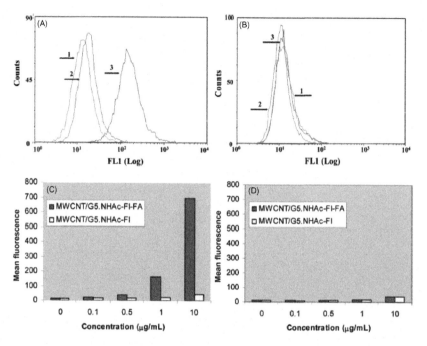

FIGURE 11.4

Flow cytometric studies of cancer cells with 1 μg/mL MWCNTs for 1 h (A and B). 1, PBS control; 2, MWCNT/G5·NHAc-FI; 3, MWCNT/G5·NHAc-FI-FA. (C and D) Dose-dependent studies.

From Shi, X., Wang, S. H., Shen, M., Antwerp, M. E., Chen, X., Li, C., et al. (2009). Multifunctional dendrimer-modified multiwalled carbon nanotubes: synthesis, characterization, and in vitro cancer cell targeting and imaging. Biomacromolecules, 1744–1750.

2. Another experiment was conducted to check the attachments of the dendrimers. Using proton nuclear magnetic resonance (^{1}H NMR) spectroscopy, a peak of 1.85 ppm represents the –CH$_3$ protons of the acetyl groups which indicate the formation of acetamide group on the dendrimers (Shi et al., 2009).

3. The MWCNTs were acetylated to remove the charge on the surface to make them neutral. To find if this was successful, a Zeta potential measurement was taken; before acetylation the surfaced potential was +28.3 mV, and after acetylation the potential was +3.1 mV (Shi et al., 2009).

The FA ligand was used as a targeting ligand and FI dye was used as an imaging molecule, providing the MWCNTs targeting and imaging capabilities. FA ligand was used because it is widely known that folic acid receptors (FARs) are common in many carcinomas such as breast, ovary, endometrium, kidney, lung, head, neck, brain, and myeloid cancers. The cancer cells used for this study were KB-HFAR and KB-LFAR. KB cancer cells are human epithelial carcinoma cells, and the KB-HFAR is high in FAR and KB-LFAR is low in FAR. KB-HFAR and KB-LFAR were treated for 1 h with 1 μg/mL MWCNT with the functionalized groups and without. KB-HFAR showed significant fluorescent signals with MWCNT/G5·NHAc-FI-FA, and KB-LFAR did not show any response to the treatment (Fig. 11.4A and B).

To study concentration dependence, the cancer cells were treated with different concentrations of MWCNTs. The uptake of the functionalized MWCNT showed a dependence on concentration for KB-HFAR with MWCNT/G5·NHAc-FI-FA, while MWCNT/G5·NHAc-FI did not respond well to the treatment. Also, KB-LFAR did not respond to the treatment, even at concentrations of 10 µg/mL (Fig. 11.4C and D). The functionalization of a drug carrier is important and in this case it was necessary for drug delivery, but the target cells must have the corresponding receptor as well (Shi et al., 2009).

11.3.4.3 Functionalized micelles for drug delivery in lung cancer

MRI-visible polymeric micelles were made to target lung cancer cells. The surface was functionalized with a lung cancer targeting peptide (LCP). The peptide is known to bind to the integrin $\alpha_v\beta_6$. A scrambled peptide (SP) was made which was the same peptide as LCP but the amino acid sequence was scrambled to be used as a control. Doxorubicin (doxo), a therapeutic drug for lung cancer and superparamagnetic iron oxide used for MRI application, was loaded into the micelle. Some of the micelles were coated with LCP and some with SP, and they were coated with different surface densities to determine the effect of surface density on cellular uptake (Guthi et al., 2009).

The lung cancer cellular uptake of the micelles was measured as a function of incubation time (Fig. 11.5A). The concentration used for the study was 19.4 µg/mL of LCP and SP. For both peptides, the cellular uptake was more rapid during the first 2 h and then slowed down and started to plateau. LCP-encoded micelles did much better than the SP micelles by approximately three times (Guthi et al., 2009).

Fig. 11.5B shows the different surface density (20% and 40%) micelles of LCP and SP that was incubated for 5 h as a function of dosage. For all the MFM, the increase in dosage increased the cytotoxicity to the cancer cells. Further analyzing the data, at the same surface density, the LCP-encoded MFM was more toxic than the SP. Similar to the other studies, it is important to note that the cellular uptake is increased if the proper ligand is coating the nanocarrier. The data also shows that surface density increases cytotoxicity. The results found in this study mimic the other research findings in terms of surface coating dependence and dosage dependence (Guthi et al., 2009).

11.4 **NANO-OPHTHALMOLOGY**

Ophthalmology refers to the study of the eye. The eye is comprised of several specialized tissues that work together to initiate visual perception in response to photons of light. Any insult to these tissues results in a consequence to vision and an impact on the quality of life for the patient. Both environmental trauma and genetic disorders can cause varying degrees of ocular diseases (Farjo, Skaggs, Alexander, & Quiambao, 2006).

Nanotechnology has many applications in disorders of eye. These include drug delivery, study of pathomechanism of eye diseases, regeneration of the optic nerve, and counteracting neovascularization involved in some degenerative disorders (Jain, 2008).

Nanotechnology has been used for ophthalmic formulations for a decade but research is still in progress to improve the delivery and safety of drugs used for treating disorders of the eye.

11.4.1 **THEORY AND BACKGROUND**

Nanotechnology has made tremendous contributions in ophthalmology. A few areas worth noting are given below.

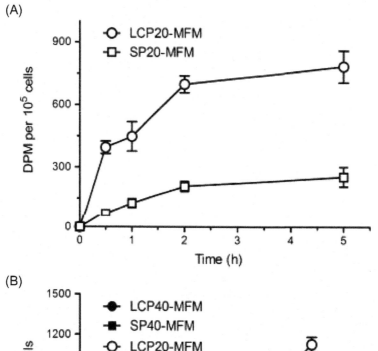

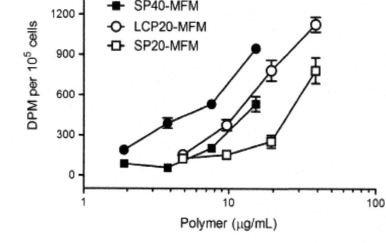

FIGURE 11.5

(A) Time-dependent uptake of multifunctional micelles (MFM) on lung cancer cells. (B) Dose-dependent uptake of the MFM on lung cancer cells.

From Guthi, J. S., Yang, S.-G., Huang, G., Li, S., Khemtong, C., Kessinger, C. W., et al. (2009). MRI-visible micellar nanomedicine for targeted drug delivery to lung cancer cells. American Chemical Society, 32–40.

11.4.1.1 Ocular drug delivery

Approximately 90% of all ophthalmic drug formulations are applied as eye drops. While eye drops are convenient, about 95% of the drug contained in the drops is lost through tear drainage, and the very tight epithelium of the cornea compromises the permeation of drug molecules. Nanocarriers, such as nanoparticles, liposomes, and dendrimers, are used to enhance ocular drug delivery. The same systems can be used to protect and release the drug in a controlled way, reducing the number of injections required. Thus, targeting the drug to the site of action leads to a decrease in the dose required and side effects (Jain, 2008).

11.4.1.2 Topical drug application to the eye

Topical application of nonsteroidal anti-inflammatory drugs on the eye is a common treatment used to treat the inflammatory reaction manifested by narrowing of the pupil induced by surgical injury such as cataract extraction. Studies have shown that polymeric nanoparticles can be used to improve the availability of sodium ibuprofen (anti-inflammatory agent) at the intraocular level (Jain, 2008).

11.4.1.3 Nanotechnology-based therapeutics for eye diseases

Nanotechnology-based therapeutics is an emerging field in which scientists from all over the world are exploring cures for many eye diseases. A significant amount of research has been carried out for the use of dendrimers for drug delivery, prevention of scar formation and for disrupting inflammation, etc. The use of nanotechnology is also being explored for prevention of neovascularization, which is a cause of many disorders related to the eye. Nanotechnology also has potential for completely regenerating the optic nerve (Jain, 2008).

11.4.2 APPLICATIONS OF NANOTECHNOLOGY IN OPHTHALMOLOGY

Below are some applications and research findings in nano-ophthalmology.

11.4.2.1 Ophthalmic drug delivery through nanoparticles in contact lenses

Scientists have invented a simple method of making polymeric lens materials that can be loaded with eye medication for ophthalmic drug delivery applications. The method of drug delivery via the lens using nanoparticles is better than the conventional drug delivery by incorporating the drug into the lens during the manufacture process (Jain, 2008).

The solution to constitute the lens contains a mixture of molecules, which create nanochannels when they set. The channels act as conduits for the drug to be released when the lens comes into contact with eye fluid. The channels also render the lens nanoporous, i.e., tears and gases can cross into and out of lens, making it more compatible with the human eye. By adjusting the channel size, medications can be delivered over hours or days (Jain, 2008).

11.4.2.2 Use of dendrimers in ophthalmology

The use of anionic dendrimers is also being investigated in ophthalmology. One of the main motives of using these compounds is to overcome the limitation of agents targeted to a single molecule or receptor. Dendrimers enable polyvalent medicines, larger molecules where several ligands can bind to several receptors in order to get the desired biological response. The use of dendrimers in drug delivery to the eye is also being explored. Dendrimers can be used to prevent scar formation following eye surgery.

Another use would be to disrupt inflammation and angiogenesis in the posterior chamber of the eye (Jain, 2008).

11.4.2.3 DNA nanoparticles for nonviral gene transfer to the eye

The eye is an excellent candidate for gene therapy as it is immune-privileged and much of the disease-causing genetics are well understood. Compacted DNA nanoparticles have been investigated as a system for nonviral gene transfer to ocular tissues. The compacted DNA nanoparticles have already been shown to be safe and effective in a human clinical trial, have no theoretical limitation on plasmid size, do not provoke immune responses, and can be highly concentrated. An experimental study has shown that DNA nanoparticles can be targeted to different tissues within the eye by varying the site of injection (Farjo et al., 2006).

11.4.2.4 DNA nanoparticles for gene therapy of retinal degenerative disorders

Research demonstrates that DNA nanoparticles corrected vision defects in a mouse model of retinitis pigmentosa by delivery of normal copies of genes into photoreceptor cells. DNA nanoparticles may also offer the potential to provide effective treatments for more complex eye disorders such as diabetic retinopathy, macular degeneration, and various diseases that injure ganglion cells and the optic nerve (Jain, 2008).

11.4.2.5 Nanobiotechnology for treatment of glaucoma

Glaucoma involves abnormally high pressure of the fluid inside the eye. However, barely 1–3% of existing glaucoma medicines penetrate into the eye. Experiments with nanoparticles have shown not only high penetration rates but also little patient discomfort (Jain, 2008).

11.4.2.6 Nanoparticle-based topical drug application to the eye

Chitosan nanoparticles: In vivo research shows that chitosan nanoparticles are promising vehicles for ocular drug delivery (Jain, 2008).

Polylactide nanoparticles: In vivo research shows that polylactide nanoparticles increase drug availability and reduce inflammation (Jain, 2008).

11.4.2.7 Nanobiotechnology for regeneration of the optic nerve

The optic nerve carries the information of vision from the eye to the brain. Fig. 11.6 shows the location of the optic nerve in the human eye (Diseases & Conditions, n.d.).

Self-assembling peptide nanofiber (SAPN) scaffolds are being used to nano-neuro knit-damaged tissue deep within the mammalian brain. After transecting the optic nerve in hamsters, scientists injected SAPN into the area, which allowed the axons in the damaged area to regenerate and knit to the surrounding neural tissue with restoration of vision. Although the technique is still far away from human use, the prospects of application are promising (Wilson, 2006).

11.5 TISSUE ENGINEERING

Advances in medical science have increased the average life expectancy. As a result, more organs, joints, and other critical body parts wear out and need to be replaced to maintain a good quality of life in old ages. The current tissue technology is expensive and considered high risk due to the possible rejection, toxicity, inflammation, and many more (Jain, 2008).

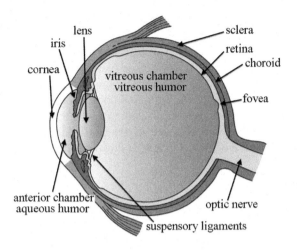

FIGURE 11.6

The human optic nerve.

From https://en.wikipedia.org/wiki/Optic_nerve#/media/File:Three_Main_Layers_of_the_Eye.png.

Tissue engineering is a fast growing scientific and technological field whose ultimate goal is to replace or restore the anatomic structure and function of damaged, injured, or missing tissue by implanting cells, scaffolds and DNA, protein or protein fragments. Therefore, engineered tissue for clinical use should resemble natural tissues both functionally and structurally, be available in variety of sizes and shapes, to continue to develop after in vivo implantation and be completely integrated into surrounding tissues. Tissue engineering is applicable in almost every part of human body such as skin, liver, heart, blood vessels, bone, bone marrow, and cartilage, to name a few (Jain, 2008).

Nanomedicine is described as the application of nanotechnology in health care and offers many promising possibilities to significantly improve medical diagnosis and therapy and in turn can lead to an affordable higher quality of life for everyone. Nanotechnology plays an integral role in creating materials and structures for tissue regeneration to match the nanoscale structures of biological cells and connective tissues with enhancement in surface properties for cell adhesion and protection (Jain, 2008).

11.5.1 THEORY AND BACKGROUND

In tissue engineering, the cells originate from three typical sources with their own disadvantages and advantages: (1) autologous is defined as cells taken from the person who is the final recipient of the new tissues, (2) allogenic is defined as cells taken from another person and transplanted into the recipient, and (3) xenogenic is defined as cells taken from another species and transplanted into the recipient. The advantage of autologous and allogenic is that they are non-immunogenic but they cannot be cultured ahead of time. On the other hand, xenogenic cells are readily available and can be harvested in large quantity but require genetic treatment before application to human can be considered (Jain, 2008).

Biomaterials for tissue engineering have been replaced with bioactive compounds capable of eliciting controlled action and reaction to stimulate and guide cell growth. These bioactive materials include glasses, ceramics, glass–ceramics, and composites as well as polymers. These materials are also the basis for tissue engineering where the purpose of the implant is to help the body to heal itself. Nanotechnology has helped material science to synthesize nanoscale macromolecular structures with finely controlled composition and architecture as well as reformation of an extracellular matrix (ECM) which provides mechanical integrity and matrix for cell attachment. Nanomaterials have also been applied as barriers to surround implanted cells and tissues to protect them from the rejection mechanisms of the host and thus allowing a wider utilization of donated organs (Jain, 2008).

Since ECM takes on different characteristics depending on the tissue, the design of scaffold and in vitro cultivation parameters, structural integrity and growth and differentiation of specific cells are very important for the progress toward bioartificial organs in this field (Jain, 2008).

11.5.2 APPLICATIONS

Tissue engineering can be applied in different parts of the human body. In this chapter, application for artificial tissues and organs will be explained in detail. This will include organ transplantation, organ-assisting device (OAD) as well as vascular scent application (Jain, 2008).

Most of tissue engineering is scaffold base; therefore, tissue engineering scaffolds should be comparable to the native ECM in terms of chemical composition and physical structure. According to Ma et al. (2005), polymeric nanofiber matrix is similar with nanoscaled nanowoven fibrous ECM protein and therefore a candidate for ECM-mimetic material. Scaffolds for tissue engineering are typically solid or porous material with isotropic characteristics and present regenerative cues but do not have the capability to guide tissue regeneration. Based on this, scientists have developed novel 3D nanofilament-based scaffolds that mimic the strategy used by fibrillar structures to guide cell migration or tissue development in a direction-sensitive manner. The advantage of this technology is that it provides directional cues for cell and tissue regeneration (Jain, 2008).

The development of effective biological scaffold materials relies on the ability to present precise environmental cues to specific cell populations to guide their position and function. This development is based on natural ECMs since it has an ordered nanoscale structure that can modulate cell behaviors critical for developmental control. According to Alsberg et al. (2006), scientists have described a method for fabricating fibrin gels with defined architecture on the nanometer scale using magnetic forces to position thrombin-coated magnetic microbeads in a defined 2D array and thereby guide the self-assembly of fibrin fibrils through catalytic cleavage of soluble fibrinogen substrate (Jain, 2008).

This magnetically guided, biologically inspired microfabrication system is unique in that large scaffolds can be formed with little starting material and therefore may be useful for future in vivo applications. In addition, nanoscale technologies can be used for controlling the features as templates for microtissue formation (Jain, 2008).

11.5.2.1 Exosomes for drug-free organ transplants

The process of exocytosis using liposomes and nanotubes has been cleverly characterized and in an effort to understand and reproduce neuronal transmission, scientists have effectively simulated natural physiology using a system composed of liposome–nanotube networks controlled by electroinjection. The system can be further controlled to investigate individual phases of this process. Several devices are used to repair, replace, or assist the function of damaged organs such as kidneys. The technologies range from those for tissue repair to those for device to take over or assist the function of the damaged organs (Jain, 2008).

Exosomes are nanovesicles shed by dendritic cells. They may hold the key to achieving the long-term acceptance of transplanted organs without the need for drugs. Exosomes are no larger than

65–100 nm; yet each contains a potent reserve of major histocompatibility complex (MHC) molecules—gene products that cells use to determine self from nonself. Because certain dendritic cells have tolerance-enhancing qualities, several approaches under study involve giving recipients donor dendritic cells that have been modified in some way (Jain, 2008).

The idea is that the modified donor cells would convince recipient cells that a transplanted organ from the same donor is not foreign. MHC-rich vesicles, siphoned from donor dendritic cells, are captured by recipient dendritic cells and processed in a manner important for cell-surface recognition. Thus one can efficiently deliver donor antigen using the exosomes as a magic bullet (Jain, 2008).

The exosomes are caught by the dendritic cells of the spleen, the site where dendritic cells typically present antigens as bounty to T cells. However, these dendritic cells internalize the exosomes instead of displaying them to T cells. Once internalized, the exosomes are ushered inside larger vesicles, special endosomes called MHC-II enriched compartments, where they are processed with the dendritic cell's own MHC molecules. This hybrid MHC-II molecule, now loaded with a peptide of donor MHC, is then expressed on the cell's surface (Jain, 2008).

Such cells are activated during chronic rejection in a process associated with the indirect pathway of immune recognition. This finding is significant because current immunosuppression therapies used in the clinical setting are not able to efficiently prevent T-cell activation via the indirect pathway. Additional research will be required to determine whether donor-derived exosomes will enhance the likelihood that an organ transplant from the same donor will be accepted (Jain, 2008).

11.5.2.2 Nanobiotechnology and OADs

OAD is an emerging area for application of nanobiotechnology which includes implants and other devices to assist or replace the impaired function of various organs. An example of this is restoration of function of the tympanic membrane of the ear by magnetically responsive nanoparticles (Jain, 2008).

Superparamagnetic iron oxide nanoparticles (SNPs) composed of magnetite ($Fe(3)O(4)$) were studied preliminarily as vehicles for therapeutic molecule delivery to the inner ear and as a middle ear implant capable of producing biomechanically relevant forces for auditory function according to Kopke et al. (2006). Magnetite SNPs were synthesized, then encapsulated in either silica or poly(D,L-lactide-co-glycolide) or obtained commercially with coatings of oleic acid or dextran (Jain, 2008).

Other technology includes nanotechnology-based human nephron filter (HNF) for renal failure. Despite the availability of various forms of renal replacement therapy for nearly four decades, mortality and morbidity are high and patients often have a poor quality of life. An HNF development could eventually enable a continuously functioning, portable, or implantable artificial kidney according to Nissenson et al. (2005).

The HNF is the first application in developing a renal replacement therapy to potentially eliminate the need for dialysis or kidney transplantation in end-stage renal disease patients. The HNF utilizes a unique membrane system created through applied nanotechnology. The ideal renal replacement device should mimic the function of natural kidneys, continuously operating, and should be adjustable to individual patient needs. No dialysis solution would be used in this device. Operating 12 h a day, 7 days a week, the filtration rate of the HNF is double that of conventional hemodialysis administered three times a week (Jain, 2008).

By eliminating dialysate and utilizing a novel membrane system, the HNF system represents a breakthrough in renal replacement therapy based on the functioning of native kidneys. The enhanced solute removal and wearable design should substantially improve patient outcomes and quality of life (Jain, 2008).

11.5.2.3 Tissue engineering and cardiology

There are several applications of nanotechnology in cardiology. An article presented by Fine, Zhang, Fenniri, and Webster (2009) introduced the design of a novel biomimetic nanostructured coating (that does not contain drugs) on conventional vascular stent materials (titanium) for improving vascular stent applications. This task was accomplished by coating the titanium surface with rosette nanotubes (RNTs). RNTs are a new class of biomimetic nanotubes that self-assemble from DNA base analogs. In this paper, the authors provided very first evidence that RNTs functionalized with lysine (RNT–K), even at low concentrations, significantly increase endothelial cell density over uncoated titanium.

The 4-h adhesion study presented in this article provided the first evidence that endothelial cells adhered at a higher density to RNT-coated titanium compared to uncoated titanium substrates. Specifically, the 0.01 and 0.001 mg/mL RNT–K-coated titanium increased endothelial cell density by 37% and 22%, respectively, over uncoated titanium as shown in Fig. 11.7 (Fine et al., 2009).

In short, this study demonstrated for the first time improved endothelial cell density on RNT-coated titanium compared to uncoated titanium controls. Importantly, this was achieved without embedding or coating drugs on the titanium stent surface, rather, such positive data was obtained after simply soaking titanium in RNT-containing solutions for 45 min. This study, thus, suggested the potential of using biomimetic RNTs as a quickly applied, efficient coating on titanium to improve cytocompatibility properties for vascular stent applications.

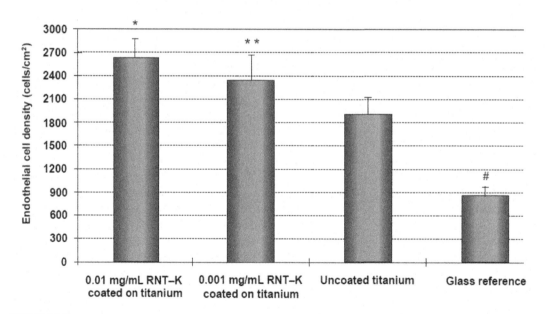

FIGURE 11.7

Enhanced endothelial cell adhesion.

*From Fine, E., Zhang, L., Fenniri, H., & Webster, T. (2009). Enhanced endothelial cell functions on rosette nanotube-coated titanium vascular stents. Data are mean values ± standard error of mean; N = 3. *p < 0.05; **p < 0.1 when compared to uncoated titanium; and #p < 0.005 when compared to all other substrates. Abbreviations: RNT–K, rosette nanotube–lysine side chain. International Journal of Nanomedicine, 91–97.*

11.6 BONE ENGINEERING

The function of human skeletal systems involves structurally supporting the body, protecting the vital organs, and serving as a reservoir of minerals for balanced metabolism. Clinically, most failures or defects of the skeletal system are induced by traumatic injuries, age-related or osteoporotic fractures, and pathological degenerations (Kuo et al., 2010).

Among the causes, age-related and osteoporotic fractures are increasingly becoming one of the major health care concerns around the world. In many cases, surgical intervention with bone grafts or even total joint replacements are needed. Another common example of bone defects is congenital deformities, with functional and cosmetic corrections of these complications becoming a major clinical practice. The surgical procedures for such purposes primarily involve the transfer of tissues or the placement of implantable prosthesis (Kuo et al., 2010).

Nowadays, numerous synthetic bone graft materials have been developed to alleviate the practical complications associated with autografts and allografts. Although good progress is being made, the function of these materials is quite different *in vivo* from that of natural bone tissues either compositionally or structurally. Recently, engineering multi-phase materials (i.e., composites) with structure and composition similar to natural bone have been attempted (Kuo et al., 2010).

Nanocomposites, particularly hydroxyapatite and collagen-based, have gained much recognition as bone grafts due to their compositional and structural similarity with natural bone. In fact, bone graft materials are not only required to be bioabsorbable and to degrade into harmless by-products that can be processed by the body, but also to stimulate new tissue generation (Jain, 2008).

11.6.1 THEORY AND BACKGROUND

The reconstructive surgery of failed or deformed skeletal systems is realized through replacing the defective tissues with viable, functioning ones. For minor fractures, bone is capable of self-regeneration within a few weeks without surgical interventions. For severe defects or deformities, bone grafting becomes necessary to restore its normal function because in this case bone cannot heal by itself. A classical approach is the transplantation of homologous bone tissue to replace the damage one. It involves the use of bone grafts to provide the defect site with initial structural stability and osteogenic environment. Autografts are bone segments taken from the patient's own body, whereas allografts are usually taken from cadaveric tissues (Kuo et al., 2010).

Although bone grafting has become a common practice in orthopedic surgeries, both autografts and allografts are limited by certain uncontrollable factors. For autografts, the major shortcoming is the limited amount of bone stock that is available for harvesting. For allografts, the major shortcoming has been the immunogenic response to the foreign tissue of the graft, in which the implant is often rejected by the body and is subject to an inflammatory reaction, transmitting diseases and terminal sterilization techniques such as gamma irradiation can compromise the mechanical integrity of the allograft (Kuo et al., 2010).

11.6.2 APPLICATION

Researchers have developed carbon nanofibers-reinforced polycarbonate urethane composite and found that such composites enhanced osteoblast function which is important for the design of successful orthopedic implants. It is very important to increase the activity of bone cells on the surface of materials used in the design of orthopedic implants so that such cells can promote integration into surrounding bone or complete replacement with naturally produced bone (Jain, 2008).

Furthermore, a nano-scale molecular scaffolding has been designed that resembles the basic structure of bone that uses the pH-induced self-assembly of a peptide-amphiphile to make a nanostructured fibrous scaffold reminiscent of ECM. Nanofibers, approximately 8 nm in size, come in the form of a gel that could be injected into a broken bone to help the fracture-mending crystallization process. The discovery is related to recreating the structure of bone at the nanoscale level and could lead to development of a hardening gel that speeds the healing of fractures. It could help patients avoid conventional surgery or be used to repair bone fractures of soldiers in battlefield and shows some features of natural bone both in main composition and hierarchical microstructure (Jain, 2008).

According to Liao et al. (2004), the 3D porous scaffold materials mimic the microstructure of cancellous bone. Cell culture and animal model tests showed that the composite material is bioactive. The osteoblasts were separated and adhered, spread, and proliferated throughout the pores of the scaffold material within a week. The scaffolds have been successfully implanted in patients in China for repair of bone defects and spinal fusion (Jain, 2008).

The nanobone material is inserted where the bone needs to heal with calcium phosphorus as critical material which is reduced to 30 nm in thickness and 60 nm in width. At this size, the properties of calcium phosphorus change. On large scale calcium phosphorus does not degrade, but on a nanoscale it does after a minimum of 6 months, and the space is filled by natural bone (Jain, 2008).

This technology is better than current methods that use ceramics or metals because those materials remain in the patient's body and can cause infection, pain, and make the repaired bone more vulnerable to fracture. The technology has been found to be effective in repairing small bones ranging from 1 to 2 cm in length, making the technology useful after removal of bone tumors (Jain, 2008).

Artificial bone scaffolds have been made from polymers or peptide fibers; however, they have low strength and possibility of rejection in the body. For this reason, single-walled carbon nanotubes (SWCNTs) have been used as scaffolds for the growth of artificial bone material according to Zhao et al. (2005). SWCNTs may lead to stronger and improved flexibility of artificial bone, new types of bone grafts and to inroads in the treatment of osteoporosis and fractures (Jain, 2008).

Another important thing in bone implant is artificial joints. Artificial joints might be improved by making the implants out of tiny carbon tubes with diameter of 60 nm and filaments that are all aligned in the same direction. This alignment is mimicking the alignment of collagen fibers and natural ceramic crystals in real bone. Besides the ability to stimulate the growth or more new bone tissue, the nanoscale materials are able to mimic the surface features of proteins and natural tissues and thus prompting the cells to stick better and cause less of rejection response from the body (Jain, 2008).

In overall, the materials used for bone implant and artificial joints need to be able to promote integration into surrounding bone or complete replacement with naturally produced bone as well as biodegradable in the body (Jain, 2008).

11.6.2.1 Antibacterial implants

Microbial colonization and biofilm formation on the surface of implant devices is a major issue and may cause peri-implantitis and lead to bone loss. Thus, an article presented by Liao, Anchun, and Quan (2010) studied the biological performance of antibacterial titanium implants. The article assessed cell viability, cytoskeletal architecture, and cell attachment on silver nanoparticle-modified titanium (Ti-nAg) plates.

It also tested the antibacterial activity of Ti-nAg and described various methods to test the biological performance of Ti-nAg such as antibacterial tests, as shown in Fig. 11.8 and anti-adhesive test and cytocompatibility assays, as shown in Fig. 11.9 (Liao et al., 2010).

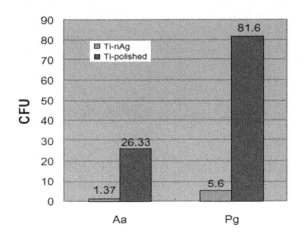

FIGURE 11.8

Inhibition of bacterial growth on Ti-nAg surface (Liao et al., 2010).

From Liao, J., Anchun, M., & Quan, Y. (2010). Antibacterial titanium plate deposited by silver nanoparticles exhibits cell compatibility. International Journal of Nanomedicine, 337–342.

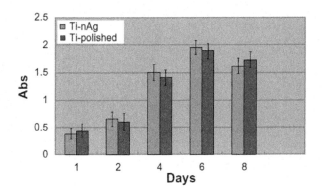

FIGURE 11.9

Cell viability estimated by the MTT assay.

From Liao, J., Anchun, M., & Quan, Y. (2010). Antibacterial titanium plate deposited by silver nanoparticles exhibits cell compatibility. International Journal of Nanomedicine, 337–342.

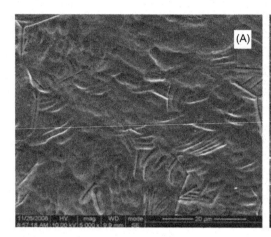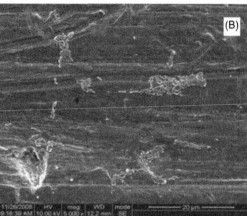

FIGURE 11.10

Scanning electron photomicrographs of the Ti-nAg (A) and Ti-polished (B) surfaces after incubation of Aa (magnification: 5000×; working distance: 8000 μm). The Aa bacteria exhibited a rod shape and reduced bacterial attachment to the Ti-nAg surface, relative to the Ti-polished surface.

From Liao, J., Anchun, M., & Quan, Y. (2010). Antibacterial titanium plate deposited by silver nanoparticles exhibits cell compatibility. International Journal of Nanomedicine, 337–342.

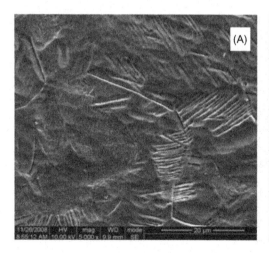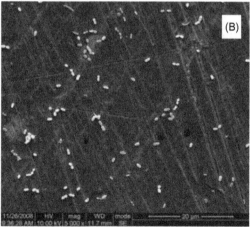

FIGURE 11.11

Scanning electron photomicrographs of the Ti-nAg (A) and Ti-polished (B) surfaces after incubation of Pg (magnification: 5000×; working distance: 8000 μm). The Pg bacteria exhibited a round shape and reduced bacterial attachment to the Ti-nAg surface, relative to the Ti-polished surface.

From Liao, J., Anchun, M., & Quan, Y. (2010). Antibacterial titanium plate deposited by silver nanoparticles exhibits cell compatibility. International Journal of Nanomedicine, 337–342.

Fig. 11.8 shows the inhibition of bacterial (types Aa and Pg) growth on Ti-nAg surface confirming the antibacterial nature of the substrate. Fig. 11.9 displays similar cell viability estimated by MTT assay which confirmed uncompromised cytocompatibility of the substrate. The anti-adhesive nature of the substrate was determined by qualitative analysis conducted with SEM as shown in Figs. 11.10 and 11.11.

Based on these figures, it can be concluded that Ti-nAg is a promising material with antibacterial properties and can be used as an implantable biomaterial.

11.7 **CONCLUSIONS**

Nanomedicine is in its early stages and many of its applications are still in the research phase. In nanopharmaceuticals, the drug carriers discussed have been successful in vitro but research still needs to be done in vivo. Nano-ophthalmology also has many applications that are in the research phase but some have reached the clinical level. Tissue engineering and bone engineering have been successful in practical applications are able to replace vital organs and bones. Nanomedicine is very promising and its future is progressing quickly. The research findings are to help the less fortunate and are paving the way to a brighter future.

References

Abelmann, L., Bos, A., & Lodder, C. (2005). *Magnetic microscopy of nanostructures*. Berlin, Heidelberg: Springer.

Aguilar, A. (2010). A hybrid nanosensor for TNT vapour detection. *Nano Letters. American Chemical Society*, 380–384.

Ahn, Y. H., Yu, G. C., Kim, Y. C., Lee, S. S., & Kim, J. (2008). Electro-optic response characteristics of a dual particle electrophoretic display system. *Molecular Crystals and Liquid Crystals*, *492*, 322–327.

Ahuja, T. (August 2010). Potentiometric urea biosensor based on multi-walled carbon nanotubes (MWCNTs)/silica composite material. *National Physical Laboratory (Council of Scientific & Industrial Research)*.

AIST. (2007). Retrieved from: http://www.aist.go.jp/index_en.html.

Alexiou, C., Arnold, W., Klein, R., & Parak, F. (2000). *Locoregional cancer treatment with magnetic drug targeting. University of Munich*.

Aliev, A. E. (2009). Giant-stroke, superelastic carbon nanotube aerogel muscles. *Materials Science*.

Andrue, T., Arbiol, J., Cabot, A., Cirera, A., Prades, J. D.Ramirez, F. H., et al. (2008). Nanosensors: controlling transduction mechanisms at the nanoscale using metal oxides and semiconductors. In F. J. Arregui (Ed.), *Sensors based on nanostructured materials* (pp. 79–83). Pamplona: Springer.

Arnall, A.H. (2003). Tools and fabrication in bottom-up manufacturing for nanotechnology. *Greenpeace report, 'Future Technologies'*.

Ashok, K. (2009). *Nanosensors for detection of biological threat contaminants in critical buildings*. Engineering Research Development Center. 18.

Aussawasathien, D., Dong, J.-H., & Dia, L. (n.d.). *Electrospun polymer nanofiber sensors*. Department of Materials and Chemical Engineering and University of Dayton Research Institute.

Averitt, R.D., Westcott, S.L., & Halas, N.J. (1998). Ultrafast electron dynamics in gold nanoshells. 16.

Barrett Research Group. (n.d.). *Introduction to liquid crystals*. Retrieved 31.03.11, from McGill University: http://barrett-group.mcgill.ca/teaching/liquid_crystal/LC03.htm.

Barroso, M. (2011). Quantum dots in cell biology. *Journal of Histochemistry and Cytochemistry*, *59*, 237–251.

Baugham et al., 2002. Carbon nanotubes—the route toward applications. doi:10.1126/science.1060928.

Belluci, S. (2005). Carbon nanotubes: physics and applications. *Physica Status Solidi*, 34–37.

Benninghoven, A. F. (1987). *Secondary ion mass spectrometry: Basic concepts, instrumental aspects, applications, and trends*. New York: Wiley.

Berndt, I., Pedersen, J. S., & Richtering, W. (2006). Temperature-sensitive core–shell microgel particles with dense shell. *Angewandte Chemie International Edition*, *45*, 1737–1741.

Bhattacharjee, A., Bhakat, D., Roy, M., & Kusz, J. (2010). Electrical conduction property of molecular magnetic material—{N(n-C4H9)4[Fe(II)Fe(III)(C2O4)3]}∞: before and after thermal degradation. *Physica B: Condensed Matter*.

Biomaterial Laboratory of the Universidad Politecnica de Madrid. (n.d.). *Overview on promising nanomaterials for industrial applications*. Madrid: Universidad Politecnica de Madrid.

Blackman, J. (2008). Metallic nanoparticles. *Handbook of metal physics*, *5*, 1–385.

Blackman, J., & Binns, C. (2008). Nanoscience and nanotechnology. *Handbook of metal physics*, *5*, 1–385.

Blum, A. S., Soto, C. M., Sapsford, K. E., & Wilson, C. D. (2010). Molecular electronics based nanosensors on a viral scaffold. *Biosensors and Bioelectronics*, *26*, 2852–2857.

Braga, P. C., & Ricci, D. (2003). *Atomic force microscopy: Biomedical methods and applications*. Humana Press.

Brinn, D. (2006). Fiat chooses Israeli startup to develop nano-sensors for measuring vehicle engine emissions. *Israel Technology*.

Bryant, D. (2001). *Liquid Crystal Institute*. Retrieved April 2011, from Advanced Liquid Crystalline Optical Materials: http://www.lci.kent.edu/switch.html.

Cameron, C. S., et al. (2002). *Advanced Functional Materials, 12,* 17.

Case Liquid Crystal Group. (2010). *Liquid Crystal Primer.* Retrieved 30.03.11, from Case Western Reserve University: http://liq-xtal.case.edu/lcdemo.htm.

Chandrashekar, S. (1992). *Liquid crystal (2nd ed.).* Cambridge University Press.

Chen, T., Ferris, R., Zhang, J., Ducker, R., & Zauscher, S. (2010). Stimulus-responsive polymer brushes on surfaces: transduction mechanisms and applications. *Progress in Polymer Science, 94–112.*

Chillcce, E., & Ramos-Gonzales, R. (2010). Luminescence of PbS quantum dots spread on the core surface of a silica microstructured optical fiber. *Journal of Non-Crystalline Solids,* 2397–2401.

Chourasia, A. R. (1997). Auger electron spectroscopy. In *Handbook of analytical chemistry.* Prentice Hall.791808 Chapter 42.

Cleland, A.N., & Roukes, M.L. (1998). A nanometre-scale mechanical electrometer. 392. Retrieved from Caltech. Edu.

Collins, P. G. (2000). Nanotubes for electronics. *Scientific American,* 67–69.

Coya, C., Álvarez, A. L., Ramos, M., de Andrés, A., Zaldo, C.Gómez, R., et al. (2008). A fluorescent stilbenoid dendrimer for solution-processed blue light emitting diodes. In *Organic Optoelectronics and Photonics III.*

Daenen, M., et al. (2003). *The wondrous world of carbon nanotubes.* Eindhoven: University of Technology.

Dahn, J., Zheng, T., Liu, Y., & Zue, J. (1995). *Science (New York, N.Y.), 270,* 590.

Das, S., Gates, A., Abdu, H., Rose, G., & Picconatto, C. (2007). Designs for ultra-tiny, special-purpose nanoelectronic circuits. *IEEE Transactions on Circuits and Systems I.*

De Miguel, J. J. (2003). *Auger electron spectroscopy.* Universidad Autónoma de Madrid. Retrieved 12.05.16, from http://www.uam.es/personal_pdi/ciencias/jdemigue/research/aes.html.

Desai, P., et al. (2009). Fabrication and characterization of gold nanostructures for cancer therapy. *International Conference on Bioencapsulation,* 2.

Diseases & Conditions. (n.d.). Retrieved April 2011, from http://health.allrefer.com/health/optic-nerve-atrophy-optic-nerve.html.

Donath, E., Sukhorukov, G.B., Caruso, F., Davis, S.A., Helmuth Möhwald, H. (1998). Novel hollow polymer shells by colloid-templated assembly of polyelectrolytes. *Angewandte Chemie International Edition.*

Dybas, C. (2008). *Scientists to assess Beijing Olympics air pollution control efforts.* National Science Foundation.

Edwards, B. C. (2000). *Design and deployment of a space elevator.* Los Alamos National Laboratory.735–737

Ellois-Behnke, R., Liang, Y. -X., You, S. -W., Tay, D., Zhang, S.So, K. -F., et al. (2006). *Proceedings of the National Academy of Sciences of the United States of America, 103,* 5054–5059.

Englund, D., Faraon, A., Fushman, I., Stoltz, N., Petroff, P., & Vuckovic, J. (2007). Generation of nonclassical states of light via photon blockade in optical nanocavities. *Nature, 450,* 857.

Eswaramoorthi, I., Sundaramurthy, V., & Dalai, A. K. (2006). Partial oxidation of methanol for hydrogen production over carbon nanotubes supported Cu-Zn catalysts. *Applied Catalysis A: General, 313,* 22–34.

Fang, Z., Fu, W., Dong, Z. D., Zhang, X., Gao, B.Guo, D., et al. (2010). Preparation and biocompatibility of electrospun poly(L-lactide-co-E-caprolactone)/fibrinogen blended nanofibrous scaffolds. *Applied Surface Science,* 4133–4138.

Farjo, R., Skaggs, J., Alexander, B., & Quiambao. (2006). *Efficient non-viral ocular gene transfer with compacted DNA nanoparticles.* Retrieved April 2011, from http://www.plosone.org/article/info%3Adoi%2F10.1371%2Fjournal.pone.0000038#s4.

Felton, M. J. (2003). On the surface with Auger electron spectroscopy. *Analytical Chemistry, 75*(11), 269–271.

Fine, E., Zhang, L., Fenniri, H., & Webster, T. (2009). Enhanced endothelial cell functions on rosette nanotube-coated titanium vascular stents. *International Journal of Nanomedicine,* 91–97.

Forrest, D.R. (2008). *Davidforrest.com.* Retrieved 11.04.11, from Some Notes on Top Down vs. Bottom Up: http://davidrforrest.com/documents/top_down_bottom_up.html.

Fredy, K. (2008). New analytical applications of gold nanoparticles. *Journal of Chemistry and Pharmacy,* 1–151.

Fushman, I., Englund, D., Faraon, A., Stoltz, N., Petroff, P., & Vuckovic, J. (2008). Controlled phase shifts with a single quantum dot. *Science (New York, N.Y.)*, *320*, 769.

Genuth, I. (2006). Buckypaper—nanotubes on steroids. *Nano Letters*.

Georgakilas, V., Pellarini, F., Prato, M., Guldi, D. M., & Melle-Franco, M. (2002). Supramolecular self-assembled fullerene nanostructures. *Proceedings of the National Academy of Sciences of the United States of America*, 5075–5080.

Giersig, M., & Hilgendorff, M. (2005). *Magnetic nanoparticle superstructures*. Germany: Wiley-VCH Verlag.

Goodwin, S. C., Bittner, C. A., Peterson, C. L., & Wong, G. (2000). *Single dose toxicity study of hepatic intra-arterial infusion of doxorubicin coupled to a novel magnetically targeted drug carrier*. University of Los Angeles Medical School.

Gordeyev, S. (2010). *Protective materials for emergency responders*. Cambridge: Institute of Nanotechnology, UK.

Gouma, P. (2010). Nanosensor and breath analyzer for ammonia. *IEEE Sensors Journal*.

Grant, J. T. (2003). *Surface analysis by Auger and X-ray photoelectron spectroscopy*. Chichester: IM Publications.

Greene, L. L. (2003). Low-temperature production of ZnO nanowire arrays. *Angewandte Chemie International Edition*, 3031–3034.

Griffiths, J. (2008). Secondary ion mass spectrometry. *Analytical Chemistry*, *80*(19), 7194–7197.

Griffiths, P. R. (1983). Fourier transform infrared spectroscopy. *Science, New Series*, *222*(4621), 297–302.

Grimm, J. (2003). Novel nanosensors for rapid analysis of telomerase activity. *Cancer*.

Guthi, J. S., Yang, S. -G., Huang, G., Li, S., Khemtong, C.Kessinger, C. W., et al. (2009). MRI-visible micellar nanomedicine for targeted drug delivery to lung cancer cells. *American Chemical Society*, 32–40.

Halas, N. (2001). The optical properties of nanoshells. *Business Week*, *4*.

Han, J. -H. (2010). Exciton antennas and concentrators from core–shell and corrugated carbon nanotube filaments of homogeneous composition. *Nature Materials*.

Heil, R. (2007, December 7). *Nanotechnology*. Retrieved 12.04.11, from Wentworth Institute of Technology: http://www.rachelheil.com/courses/Nanotechnology/Nanosensors.pdf.

Hendrych, A., Kubinek, R., & Zhukov, A. V. (2007). The magnetic force microscopy and its capability for nano-magnetic studies—the short compendium. *Modern Research and Educational Topics in Microscopy*.

Heng-Yong, N. (2010, November 5). *Scanning probe microscopy*. Retrieved 01.04.11, from University of Western Ontario: http://publish.uwo.ca/~hnie/spmman.html#top.

Henry, T. (2008, November 6). *Samsung Electronics demonstrates first color large scale EPD e-paper*. Retrieved 12.04.10, from Printed Electronics World: http://www.printedelectronicsworld.com/articles/samsung_electronics_demonstrates_first_color_large_scale_epd_e_paper_00001121.asp?sessionid=1.

Hirlekar, R., et al. (2009). Carbon nanotubes and its applications: a review. *Asian Journal of Pharmaceutical and Clinical Research*, 1–11.

Hirsch, L.R., Sershen, S.R., Halas, N. J., Stafford, R. J., Hazle, J., West, J. L. (2003). Nanoshell mediated near infrared photothermal tumor therapy. *Department of Bioengineering*, *4*.

Hirst, A., Escuder, B., Miravet, J., & Smith, D. (2008). High-tech applications of self-assembling supramolecular nanostructured gel-phase materials: from regenerative medicine to electronic devices. *Angewandte Chemie International Edition*, *47*, 8002–8018.

Holister, P., Harper, T. E., & Vas, C. R. (2003). Nanotubes white paper. *CMP Cientifica*, 5–7.

Huefner, S. (2006, November 6). *Chem.usu*. Retrieved 10.04.11, from USU: www.chem.usu.edu/~tapaskar/Sara.

Huei, J. (2005, December 5). *Headlines at Hopkins*. Retrieved 10.04.11, from http://www.jhu.edu/news/home05/dec05/dnanano.html.

Ildiko, C., Genoveva, F., & Miklos, Z. (2006). Smart nanocomposite polymer membranes with on/off switching control. *Macromolecules*, 1939–1942.

ISO. (2010). Retrieved 16.03.10, from http://www.iso.org/iso/iso_technical_committee.html?commid=381983.

Jain, K. K. (2008). *Handbook of nanomedicine*. Totowa, NJ: Humana Press.

Jensen, K. (2007). Nanotube radio. *Nano Letters*, 3508–3511.

Jensen, K., et al. (2007). Buckling and kinking force measurements on individual multiwalled carbon nanotubes. *Physical Review B, 76*.

Jia, G.L. (2006, June 8). *Electrically controlled nanowire-based chemical sensors*. Retrieved 13.04.11, from SPIE: http://spie.org/x8765.xml?ArticleID=x8765.

Jin, X., Zhang, X.Wu, Z., et al. (2009). Amphiphilic random glycopolymer based on phenylboronic acid: synthesis, characterization, and potential as glucose-sensitive matrix. *Biomacromolecules*.

Johnston, H. (2008, July 21). *Nanotube cantilever weigh up*. Retrieved 10.04.11, from Physicsworld: http://physicsworld.com/cws/article/news/35081.

Kalbacova, (2008). TiO_2 nanotubes: photocatalyst for cancer cell killing. *Physica Status Solidi*, 194–196.

Kalele, S., Gosavi, S., Urban, J., & Kulkarni, K. (2006). Nanoshell particles: synthesis, properties and applications. *Current Science, 15*.

Karkare, M. (2008). *Nanotechnology: fundamentals and applications*. New Delhi: I. K. International.

Keller, N., Pham-Huu, C., Estournès, C., Grenèche, J. -M., Ehret, G., & Ledoux, M. J. (2004). Carbon nanotubes as a template for mild synthesis of magnetic $CoFe_2O_4$ nanowires. *Carbon, 42*, 1395–1399.

Kelsall, R., Hamley, I., & Geoghegan, M. (2005). *Nanoscale science and technology*. Hoboken: John Wiley & Sons Ltd.

Kevin, C. (2010). Nanosensors and nanomaterials for monitoring glucose in diabetes. *Trends in Molecular Medicine, 16*.

Kimble, H. J. (2008). Review article: the quantum internet. *Nature, 453*, 1023.

Kinloch, I., Li, Y., & Windle, A. (2004). Direct spinning of carbon nanotube fibers from chemical vapor deposition synthesis. *Science Magazine*, 276–278.

Kohler, T., Mietke, S., Ilgner, J., & Werner, M. (2003). *Nanotechnology-markets & trends*.

Kuang, Y., & Walt, D. R. (2007). Detecting oxygen consumption in the proximity of *Saccharomyces cerevisiae* cells using self-assembled fluorescent nanosensors. *Biotechnology and Bioengineering, 96*(2), 318–325.

Larminie, J. (2003). *Fuel cell systems explained*. SAE International.

Lee, W. -E., Oh, C. -J., Kang, I. -K., & Kwak, G. (2010). Diphenylacetylene polymer nanofiber mats fabricated by freeze drying: preparation and application for explosive sensors. *Macromolecular Journals*.

Lewinsky, A. A. (2007). *Hazardous materials and wastewater: Treatment, removal and analysis*. New York: Nova Science Publishers Inc.

Lewis, J.B. (n.d.). *A short history of nanotechnology*. Retrieved 13.04.11, from Foresight Institute: http://www.foresight.org/nano/history.html.

Li, D., He, Q., & Li, J. (2009). Smart core/shell nanocomposites: intelligent polymers modified gold nanoparticles. *Advances in Colloid and Interface Science*, 28–38.

Li, M., & Keller, P. (2009). Stimuli-responsive polymer vesicles. *Soft Matter*, 927–937.

Liao, J., Anchun, M., & Quan, Y. (2010). Antibacterial titanium plate deposited by silver nanoparticles exhibits cell compatibility. *International Journal of Nanomedicine*, 337–342.

Likharev, K. (1999). Single-electron devices and their applications. *Proceedings of the IEEE, 87*, 606–632. *April*.

Lin, Y. -F. (2009). *Nanosensors for monitoring water quantity and quality in public water systems*. Chicago: Champaign.

Link, J. R., & Sailor, M. J. (2003). Smart dust: self-assembling, self-orienting photonic crystals of porous Si. *Proceedings of the National Academy of Sciences of the United States of America*, 10607–10610.

Liu, T. -Y., Hu, S. -H.Liu, T. -Y., et al. (2006). Magnetic-sensitive behavior of intelligent ferrogels for controlled release of drug. *Langmuir*.

Loo, C., Lowery, A., Halas, N., West, J., & Drezek, R. (2005). Immunotargeted nanoshells for integrated cancer imaging and therapy. In *Departments of Bioengineering and Electrical and Computer Engineering*. Texas: Rice University.

Lu, A. -H., Salabas, E., & Schth, F. (2007). *Magnetic nanoparticles: synthesis, protection, functionalization, and application*. Weinheim: Verlag GmbH & Co. KGaA.

Maeda-Mamiya, R., Noiri, E., Isobe, H., Nakanishi, W., Okamoto, K., Doi, K., et al. (2009). In vivo gene delivery by cationic tetraamino fullerene.

Malik, P., Bubnov, A. M., & Raina, K. K. (2008). Electro-optic and thermo-optic properties of phase separated polymer dispersed liquid crystal films. *Molecular Crystals and Liquid Crystals, 494*, 242–251.

Malsch, N. H. (2005). *Biomedical nanotechnology*. Boca Raton: Taylor & Francis Group.

Mansoori, G. A. (2005). *Principles of nanotechnology*. New Jersey: World Scientific Publishing Co. Pte, Ltd.

Martin-Palma, R. J., & Lakhtakia, A. (2010). *Nanotechnology: A crash course*. Washington: SPIE.

McCluskey, W. M. (2005). Infrared spectroscopy of impurities in ZnO nanoparticles. *Materials Research Society*.

Min-Feng, (2000). Strength and breaking mechanism of multiwalled carbon nanotubes under tensile load. *Science (New York, N.Y.)*, 637–640.

Modir, A. (2010). *Fundamental aspects of biosensor fabrication and functionality*. Department of Chemistry, Memorial University of Newfoundland.

Moore, R. (2009, December 14). *Nanomedicine, why is it different?* Retrieved 12.04.11, from Institute of Nano Technology: http://www.nano.org.uk/articles/25.

Motomov, M., Roiter, Y., Tokarev, I., & Minko, S. (2010). Stimuli-responsive nanoparticles, nanogels and capsules for integrated multifunctional intelligent systems. *Progress in Polymer Science*, 174–211.

Nan, A., Bai, X., Son, S. J., Lee, S. B., & Ghandehari, H. (2008). Cellular uptake and cytotoxicity of silica nanotubes. *Nano Letters*, 2150–2154.

Nano-Tech-Views. (n.d.). *Traffic monitoring sensor with nanotechnology*. Retrieved 15.04.11, from Nano-Tech-Views: http://www.neno-tech-views.com/?p=667.

NASA Science. (2000, September 7). Retrieved 12.04.10, from Audacious & Outrageous: Space Elevators: http://science.nasa.gov/science-news/science-at-nasa/2000/ast07sep_1/.

Nguyen, T. -H., Kin, Y. -H., Song, H. -Y., & Lee, B. -T. (2010). Nano Ag loaded PVA nano-fibrous mats for skin application. *Journal of Biomedical Materials Research B*, 225–233.

Nicolet, T. (2001). *Introduction to Fourier transform infrared spectrometry*. Thermo Nicolet Corporation.

Nirmala, R., Park, H. -M., Navamathavan, R., Kang, H. -S., El-Newehy, M., & Kim, K. Y. (2010). Lecithin blended polyamide-6 high aspect ratio nanofiber scaffolds via electrospinning for human osteoblast cell culture. *Material Science and Engineering C*, 486–493.

Nishio, K., Ikeda, M., Gokon, N., Tsubouchi, S., & Na, H. (2007). Preparation of size-controlled (30–100 nm) magnetite nanoparticles for biomedical applications. *Journal of Magnetism and Magnetic Materials, 310*, 2408–2410.

No, Y. -S., & Jeon, C. -W. (2009). Effect of alignment layer on electro-optic properties of polymer-dispersed liquid crystal displays. *Molecular Crystals and Liquid Crystals, 513*, 98–105.

Nobel Media A.B. (2014). "Richard P. Feynman—Biographical". Retrieved 12.05.16, from *Nobelprize.org*: http://www.nobelprize.org/nobel_prizes/physics/laureates/1965/feynman-bio.html.

Novak, M., Jager, C., Kropp, H., Clark, T., Halik, M. (2009) The morphology of integrated self-assembled monolayers and their impact on devices—a computational and experimental approach.

Novotny, L., & Hect, B. (2006). *Principles of nano-optics*. Cambridge University Press.

Nowak, B., Krug, H., & Height, M. (2011). *120 Years of nanosilver history: implications for policy*.

Oldenburg, S. J., et al. (1999). Infrared extinction properties of gold nanoshells. *Applied Physics Letters, 3*.

Oltean, M. (2006). Switchable glass: a possible medium for evolvable hardware. In *NASA conference on adaptive hardware systems*. IEEE CS Press. 81–87

Organic and Biomolecular Chemistry. (2009). Retrieved 31.03.11, from Royal Society of Chemistry: http://www.rsc.org/publishing/journals/ob/article.asp?Type=Issue&Journalcode=OB&Issue=16&SubYear=2009&Volume=7&Page=0&GA=on.

Pankhurst, Q., Connolly, J., Jones, S., & Dobson, J. (2003). Application of magnetic nanoparticles in biomedicine. *Journal of Physics D: Applied Physics*.

Park, S., Kim, H., Lim, H., & Kim, C. O. (2008). Surface-modified magnetic nanoparticles with lecithin for applications in biomedicine. *Current Applied Physics, 8*, 706–709.

Patololsky, F. Z. (2006). Nanowire sensors for medicine and life sciences. *Nanomedicine: Nanotechnology, Biology, and Medicine, 1*(1), 51–65. Retrieved 10.04.11, from Harvard University—Department of Chemistry and Chemical Biology, and Division of Engineering and Applied Sciences: http://echinacea.harvard.edu/assets/Nanomedicine_1_51.pdf.

Patolsky, F. L. (2005). Nanowire nanosensors. *Materials Today April 2005*.

Peng, G. (2010). Detection of lung, breast, colorectal, and prostate cancers from exhaled breath using a sing array of nanosensors. *British Journal of Cancer*, 542–551.

Peng, S., O'Keeffe, J., Cho, C.W., Kong, J., Chen, R., & Dai, N.F. (2001). Carbon nanotube chemical and mechanical sensors. *Conference paper for the 3rd international workshop on structural health monitoring*, 1–8.

Perez, J. M. (2008). Integrated nanosensors to determine levels and functional activity of human telomerase. *Neoplasia*.

Pichot, C. (2004). Surface-functionalized latexes for biotechnological applications. *Current Opinion in Colloid & Interface Science*, 213–221.

Pichot, C., Elaissari, A., Duracher, D., Meunier, F., & Sauzedde, F. (2001). Hydrophillic stimuli-responsive particles for biomedical applications. *Macromolecular Symposia*, 285–297.

Polymer-Nanoparticle Composites Part 1 (Nanotechnology). (2010, May 25). Retrieved 13.04.11, from What-When-how In depth information: http://what-when-how.com/nanoscience-and-nanotechnology/polymer-nanoparticle-composites-part-1-nanotechnology/.

Poole, C. P., Jr, & Owens, F. (2003). *Introduction to nanotechnology*. Wiley.

Pop, E., et al. (2005). Thermal conductance of an individual single-wall carbon nanotube above room temperature. *Nano Letters*, 96–100.

Popov, M., et al. (2002). Superhard phase composed of single-wall carbon nanotubes. *Physical Review B, 65*.

Postma, H. W., Teepen, T., Yao, Z., & Grifoni, M. (2001). Carbon nanotube single-electron transistors at room temperature. *Science (New York, N.Y.)*, 76–79.

Pradeep, T. (2009). *Nano the essentials: Understanding nanoscience and nanotechnology*. New Delhi: Tata McGraw-Hill Publishing Limited Company.

Radloff, C., & Vaia, R. (2005). Metal nanoshell assembly on a virus bioscaffold. *Nano Letters, 5*.

Radtchenko, I., Sukhorukov, G., & Mohwald, H. (2007). Incorporation of macromolecules into polyelectrolyte micro- and nanocapsules via surface controlled precipitation on colloidal particles. *Colloids and Surfaces A*, 127–133.

Rahman, A., Won, M. -S., Kwon, N. -H., Yoon, J. -H., Park, D. -S., & Shim, Y. -B. (2008). *Water sensor for a nonaqueous solvent with poly(1,5-diaminonapthalene) nanofibers*. Department of Chemistry and Center for Innovative BioPhysio Sensor Technology.

Raschke, G., et al (2004). Gold nanoshells improve single nanoparticle molecular sensors. *Nano Letters, 5*.

Riboh, J. H. (2003). A nanoscale optical biosensor: real-time immunoassay in physiological buffer enabled by improved nanoparticle adhesion. *The Journal of Physical Chemistry B, 107*, 1772–1780.

Richards, R. (2009). *Introduction to nanoscale materials in chemistry*. John Wiley & Sons.

Riu, J. M. (2005, April 26). Nanosensors in environmental analysis. *Elsevier, 69*(2), 288–301.

Roca, A., Costo, R., Rebolledo, F., & Veintemillas, S. (2009). Progress in the preparation of magnetic nanoparticles for applications in biomedicine. *Journal of Physics D: Applied Physics, 42*, 11.

Roco, M. (1999). *Nanotechnology research directions: Vision for the next decade*.

Roco, M. (2010). *The long view of nanotechnology development: The National Nanotechnology Initiative at 10 years*.

Roco, M. (2011). Nanotechnology research directions for societal needs in 2020: summary of international study. *Journal of Nanoparticle Research*.

Roiter, Y., & Minko, S. (2005). Single molecule experiments at the solid-liquid interface: in situ conformation of absorbed flexible polyelectrolyte chains. *Journal of the American Chemical Society*, 15688–15689.

Roy, D., Shastri, B., Imamuddin, M., Mukhopadhyay, K., & Bhasker, K. U. (2010). Nanostructured carbon and polymer materials—synthesis and their applications in energy conversion devices. *Renewable Energy*, 1014–1018.

Ruhe, J., Ballauff, M., Biesalski, M., Dziezok, P., Grohn, F., & Johannsmann, D. (2004). Polyelectrolyte brushes. *Advances in Polymer Science*, 79–150.

Ruoff, R., et al. (2003). Mechanical properties of carbon nanotubes: theoretical predictions and experimental measurements. *Comptes Rendus Physique*, *4*, 993–1008.

Sahoo, D. R., Argawal, P., & Salapaka, M. M. (2007). Transient force atomic force microscopy: a new nano-interrogation method. In *American Control Conference*. New York: Iowa State University.

Saion, S., et al. (2005). Off-axis thermal properties of carbon nanotube films. *Journal of Nanoparticle Research*, *7*, 651–657.

Samuel, D. (2011). The role of nanotechnology in diabetes treatment: current and future perspectives. *International Journal of Nanotechnology*, 53–65.

Sarma, H. (2003). Metal nanoparticles: applications in molecular nanotechnology and DNA chip detection. *Journal of Biomolecular Structure and Dynamics*, 17–21.

Schmaljohann, D. (2006). *Thermo- and pH-responsive polymers in drug delivery*. Elsevier.1655–1670

Schmid, G. (2006). *Nanoparticles: From theory to application*. London: Wiley VCH.

Schmidt. (2006). *Nanotechnology: Assessment and perspectives*. Berlin: Springer-Verlag.

Sengupta, R. (2009). Novel nano-sensor for biomedical and industrial applications. *Banpil Photonics*.

Shabani, I., Hasani-Sadrabadi, M. M., Haddadi-Asl, V., & Soleimani, M. (2010). Nanofiber-based polyelectrolytes as novel membranes for fuel cell applications. *Journal of Membrane Science*, 233–240.

Shah, M. A. (2008). Formation of zinc oxide nanoparticles by the reaction of zinc metal with methanol at very low temperature. In *African Physics* 211.

Shah, M. A., & Ahmad, T. (2010). *). Principles of nanoscience and nanotechnology*. Alpha Science International.

Shchukin, D. G., & Sukhorukov, G. B. (2005). Halloysite nanotubes as biomimetic nanoreactors. *Small (Weinheim an der Bergstrasse, Germany)*, *1*, 510–513.

Shelley, S. (2008). Nanosensors: evolution, not revolution… yet. *CEP*, 1–5.

Shi, X., Wang, S. H., Shen, M., Antwerp, M. E., Chen, X.Li, C., et al. (2009). Multifunctional dendrimer-modified multiwalled carbon nanotubes: synthesis, characterization, and in vitro cancer cell targeting and imaging. *Biomacromolecules*, 1744–1750.

Sinha, N. (2005). Carbon nanotubes for biomedical applications. *IEEE Transactions on Nanobioscience*, *4*(2), 180–195. *June*.

Skaff, H., Emrick, T., & Rotello, V. (2008). *Nanoparticles building blocks for nanotechnology*. Amherst.

Skaraupo, S. (2007, June). *A nanotechnology test system*. Retrieved 10.04.11, from Evaluation Engineering: http://evaluationengineering.com/features/0607/0607nanoelectronics.asp.

Smith, A. W. (2010). Size-minimized quantum dots for molecular and cellular imaging. *Chemical Physics*, *96* (Part 4), 187–201.

Song. (2010). Arrays of sealed silicon nanotubes as anodes for lithium ion batteries.

Sridevi, D., & Rajendran, K. (2009). Preparation of ZnO nanoparticles and nanorods by using CTAB assisted hydrothermal method. *International Journal of Nanotechnology and Applications*, 43–48.

Srinivasan, (2007). Scanning electron microscopy of nanoscale chemical patterns. *American Chemical Society*, 191–201.

Stanacevic, M. (2004). Nanosensor device for breath acetone detection. *Sensor*.

Steff. (2010, March). *File:Schema MEB (en).svg*. Retrieved 17.04.11, from Wikipedia: http://en.wikipedia.org/wiki/File:Schema_MEB_%28en%29.svg.

Swamy, R., Rajagopalan, P., Vippa, P., Thakur, M., & Sen, A. (2007). Quadratic electro-optic effect in a nano-optical material based on the non-conjugated conductive polymer, poly(ethylenepyrolediyl) derivative. *Solid State Communications*, 519–521.

Teng, M., Wang, H., Li, F., & Zhang, B. (2010). Thio-ether-functionalized mesoporous fiber membranes: sol-gel combined electrospun fabrication and their applications for Hg^{2+} removal. *Journal of Colloid and Interface Science*, 23–28.

Thomas, O., Loubens, A., Gergaud, P., & Labat, S. (2006). X-ray scattering: a powerful probe of lattice strain in materials with small dimensions. *Science Direct*, 182–187.

Tisch, U. (2010). Diagnosing lung cancer in exhaled breath using gold nanoparticles. *Nature Nanotechnology*, 669–673.

Tsang, S., Yu, C., Tang, H., He, H., Castelleto, V.Hamley, I., et al. (2008). Assembly of centimeter long silica coated FePt colloid crystals with tailored interstices by magnetic crystallization. *Chemistry of Materials*, 4554–4556.

UnderstandingNano. (2007). *Chemical sensors and nanotechnology*. Retrieved 10.04.11, from http://www.understandingnano.com/sensor.html.

Van de Voort, F. (2009). *Quantitative FTIR condition monitoring—analytical wave of the future*. McGill IR Group, McGill University.

Varadan, V. K., Chen, L., & Xie, J. (2008). *Nanomedicine: Design and applications of magnetic nanomaterials, nanosensors and nanosystems*. Chichester: Wiley.

Vaskecich, A. K. (2008). *Localized surface plasmon (LSPR) spectroscopy in biosensing*. John Wiley & Sons.

Vazquez, M., Asenjo, A., Morales, M. D., Pirota, K. R., Confalonieri, G. B., & Velez, M. H. (2008). Nanostructured magnetic sensors. In F. J. Arregui (Ed.), *Sensors based on nanostructured materials* (pp. 196–198). Pamplona, Spain: Springer.

Vlasov, Y., Green, W. M. J., & Xia, F. (2008). Integrated GHz silicon photonic interconnect with micrometer-scale modulators and detectors. *Nature Photonics*, 2, 242.

Wang, A. T., Cui, Y., Duan, L., & Yang, Y. L. (2002). Assembly of environmental sensitive microcapsules of PNIPAAm and alginate acid and their application in drug release. *Journal of Colloid and Interface Science*, 127–133.

Wang, M., & Gates, B. (2009, May). Directed assembly of nanowires. *Materials Today*, 12(5), 34–43.

Wang, X., Li, Q., Xie, J., Jin, Z., Wang, J.Li, Y., et al. (2009). Fabrication of ultralong and electrically uniform single-walled carbon nanotubes on clean substrates. *Nano Letters*, 3137–3141.

Wang, X., Shen, Y., Xie, A., Li, S., Cai, Y.Wang, Y., et al. (2011). Assembly of dandelion-like Au/PANI nanocomposites and their application as SERS nanosensors. *Biosensors and Bioelectronics*, 26, 1–5.

Wei, J. (2004). Carbon nanotube filaments in household light bulbs. *Applied Physics Letters*.

Whitesides, G. M. (1995, November 11). *Self-assembly and nanotechnology*. Harvard University. Retrieved 10.04.11, from Fourth Foresight Conference on Molecular Nanotechnology: http://www.zyvex.com/nanotech/nano4/whitesidesAbstract.html.

Wiesendanger, R. (1995). Future nanosensors. In W. Gopel, J. Hesse, & J. N. Zemel (Eds.). *Sensors: A comprehensive survey* (vol. 8, pp. 338–353). New York, Weinheim: Wiley.

Wikipedia. (n.d.). *Wikipedia*. Retrieved 13.0.4.11, from Scanning tunneling microscope: http://en.wikipedia.org/wiki/Scanning_tunneling_microscope.

Wikipedia. (n.d.). *Wikipedia*. Retrieved 14.04.11, from http://en.wikipedia.org/wiki/Nanotechnology.

Wikipedia. (n.d.). *Wikipedia*. Retrieved 12.05.16, from Carbon nanotube: <https://en.wikipedia.org/wiki/Carbon_nanotube>.

Wilner, O. I., & Henning, A. (2010). Covalently linked DNA nanotubes. *American Chemical Society*, 1458–1465.

Wilson, B. (2006, September). *Email Address*. Retrieved April 2011, from http://www.dddmag.com/nano-neuro-knitting.aspx.

Wu, W., Aiello, M., Zhou, T., Berliner, A., Banerjee, P., & Zhou, S. (2010). In-situ immobilization of quantum dots in polysaccharide-based nanogels for integration of optical pH-sensing, tumor cell imaging, and drug delivery. *Biomaterials*.

Xiao, L. (2008). *Flexible, stretchable, transparent carbon nanotube thin film loudspeakers. ACS Publications*.

Xiao, Y. (October 2009). *Anti-HER2 IgY antibody-functionalized single-walled carbon nanotubes for detection and selective destruction of breast cancer cells*. Chemical Science and Technology Laboratory, National Institute of Standards and Technology (NIST).

Xiong, S., Yuan, C., Zhang, X., & Qian, Y. (2011). Mesoporous NiO with various hierarchical nanostructures by quasi-nanotubes/nanowires/nanorods self-assembly: controllable preparation and application in supercapacitors. *CrsytEngComm, 13*, 626–632.

Xu, K., Cao, P., & Heath, J. R. (2009). Scanning tunneling microscopy characterization of the electrical properties. *Nano Letters*.

Yang, R., Gu, Y., Li, Y., Zheng, J., & Li, X. (2010). Self-assembled 3-D flower-shaped SnO_2 nanostructures with improved electrochemical performance for lithium storage. *Acta Materialia, 58*, 866–874.

Yao, B., Heinrich, H., Smith, C., Bergh, M. V., Cho, K., & Sohn, Y. -H. (2011). Hollow-cone dark-field transmission electron microscopy for dislocation density characterization of trimodal Al composites. *Micron (Oxford, England: 1993)*, 29–35.

Yu, H., Oduro, W., Tam, k, & Tsang, E. (2008). Chemical methods for preparation of nanoparticles in solution. *Handbook of metal physics, 5*, 1–385.

Yusa, S., Yamago, S., Sugahara, M., Morikawa, S., Yamamoto, T., & Morishima, Y. (2007). *Macromolecules*, 5907.

Zhang, D., & Lin, C. (n.d.). *Carbon nanotubes: final report*.

Zhang, P., Le, K., Malalur-Nagaraja-Rao, S., Hsu, L., & Chiao, J. (2005). *Self-assembly micro optical filters*. Society of Photographic Instrumentation Engineers.

Zhang, Y. W., Wang, Z., Wang, Y., Zhao, H., & Wu, C. (2007). Facile preparation of pH-responsive gelatin-based-core-shell polymeric nanoparticles at high concentrations via template polymerization. *Polymer*, 5639–5645.

Zhao, J. B. (2002). Gas molecule adsorption in carbon nanotubes and nanotube bundles. *Nanotechnology, 13*(2002), 195–200.

Zhong, L., Hu, J., Liang, H., Cao, A., & Song, W. (2006). Self-assembled 3D flowerlike iron oxide nanostructures and their application in water treatment. *Advanced Materials*, 2426–2431.

Zhou, W., & Wang, Z. L. (2006). *Scanning microscopy for nanotechnology*. New Orleans: Springer.

Zhuang Liu, K. C. (August 12, 2008). Drug delivery with carbon nanotubes for in vivo cancer treatment. In *The Molecular Imaging Program at Stanford; Department of Chemistry, Stanford University*.

FURTHER READING

Advincula, R., Brittain, W., Caster, K., & Ruhe, J. (2004). *Polymer brushes: Synthesis, characterization, applications*. New Jersey: Wiley.

Batteries, Supercapacitors and Fuel Cells: Scope. (2007). *Science References Services*.

Blum, A.S. (2010). Molecular electronics based nanosensors on a viral scaffold. 26.

Carbon Nanotubes. (2009). Retrieved from Nanocyl: The carbon nanotube specialist: http://www.nanocyl.com/CNT-Expertise-Centre/Carbon-Nanotubes.

Cellulose. (n.d.). Retrieved from Wikipedia: http://en.wikipedia.org/wiki/Cellulose.

Emissions, J.M.-E. (2010, March 22). *Carbon nanotube-based batteries for HEVs*. Retrieved 10.04.11, from http://www.justmeans.com/Carbon-Nanotube-based-Batteries-for-HEVs/11428.html.

Gel. (n.d.). Retrieved from Wikipedia: http://en.wikipedia.org/wiki/Gel.

Haghi, A. K. (2009). *Electrospun nanofibers research: recent developments*. New York: Nova Science Publishers, Inc.

N/A. (2011). *Introduction to structures and solids*. Retrieved 17.04.11, from Science is Cool: http://www.scienceiscool.org/solids/intro.html#xray.

Park, M. S., Kang, Y. M., Wang, G. X., Dou, S. X., & Liu, H. K. (2008). *Advanced Functional Materials, 15*, 1845.

Polymer Brush. (n.d.). Retrieved from Wikipedia: http://en.wikipedia.org/wiki/Polymer_brush.

Pugnetti, S., Dolcini, F., & Fazio, R. (2007). Dc Josephson effect in metallic single-walled carbon nanotubes (cond-mat/0702678). *Solid State Communications*, 551–556.

Wang, D., Choi, D., Li, J., Yang, Z., Nie, Z.Kou, R., et al. (2009). Self-assembled TiO_2-graphene hybrid nanostructures for enhanced Li-ion insertion. In *American Chemical Society* 907–914.

Wang, X.S. (2011). Assembly of dandelion-like Au/PANI nanocomposites and their application as SERS nanosensors.

West, J. (2005). Immunotargeted nanoshell for integrated cancer imaging and therapy. *Nano Letters, 5*.

Younge-Ju, K., Ju Young, C., Heon, H., Hoon, H., Dae-Sup, S., & Inpil, K. (2010). Preparation of piezoresistive nano smart hybrid material based on grapheme. *Elsevier: Current Applied Physics*, 350–352.

Zhang, Y., Ming, M.A., Ning, G.U., Ling, X.U., & Kun, J. (2004). Preparation of silver nanoshells on silica particles by a simple two-step process. Chinese Chemical Letters, 3.

Index

Note: Page numbers followed by "*f*" and "*t*" refer to figures and tables, respectively.

Printed in the United States
By Bookmasters